Veterinary Medicines in the Environment

T0174259

Other Titles from the Society of Environmental Toxicology and Chemistry (SETAC)

For information about SETAC publications, including SETAC's international journals, Environmental Toxicology and Chemistry and Integrated Environmental Assessment and Management, contact the SETAC Administratice Office nearest you:

SETAC Office
1010 North 12th Avenue
Pensacola, FL 32501-3367 USA
T 850 469 1500   F 850 469 9778
E setac@setac.org

SETAC Office
Avenue de la Toison d'Or 67
B-1060 Brussells, Belguim
T 32 2 772 72 81   F 32 2 770 53 86
E setac@setaceu.org

www.setac.org
Environmental Quality Through Science®

Veterinary Medicines in the Environment

Edited by

Mark Crane
Alistair B. A. Boxall
Katie Barrett

From the SETAC Pellston Workshop on
Veterinary Medicines in the Environment
Pensacola, Florida, USA
12–16 February 2006

Coordinating Editor of SETAC Books
Joseph W. Gorsuch
Gorsuch Environmental Management Services, Inc.
Webster, New York, USA

SETAC

CRC Press
Taylor & Francis Group
Boca Raton London New York

CRC Press is an imprint of the
Taylor & Francis Group, an **informa** business

CRC Press
Taylor & Francis Group
6000 Broken Sound Parkway NW, Suite 300
Boca Raton, FL 33487-2742

First issued in paperback 2019

© 2009 by Taylor & Francis Group, LLC
CRC Press is an imprint of Taylor & Francis Group, an Informa business

No claim to original U.S. Government works

ISBN-13: 978-1-4200-8424-5 (hbk)
ISBN-13: 978-0-367-38685-6 (pbk)
ISBN: 978-1-880611-94-4 (SETAC Press)

Library of Congress Cataloging-in-Publication Data

Veterinary medicines in the environment / editors, Mark Crane, Alistair B.A. Boxall, Katie Barrett.
    p. cm.
    Includes bibliographical references.
    ISBN 978-1-4200-8424-5 (alk. paper)
    1. Veterinary drugs--Environmental aspects. I. Crane, Mark, 1962- II. Boxall, Alistair B. A. III. Barrett, Katie. IV. Title.

SF917.E33 2008
628.5'2--dc22
    2008019579

**Visit the Taylor & Francis Web site at
http://www.taylorandfrancis.com**

**and the CRC Press Web site at
http://www.crcpress.com**

SETAC Publications

Books published by the Society of Environmental Toxicology and Chemistry (SETAC) provide in-depth reviews and critical appraisals on scientific subjects relevant to understanding the impacts of chemicals and technology on the environment. The books explore topics reviewed and recommended by the Publications Advisory Council and approved by the SETAC North America, Latin America, or Asia/Pacific Board of Directors; the SETAC Europe Council; or the SETAC World Council for their importance, timeliness, and contribution to multidisciplinary approaches to solving environmental problems. The diversity and breadth of subjects covered in the series reflect the wide range of disciplines encompassed by environmental toxicology, environmental chemistry, and hazard and risk assessment, and life-cycle assessment. SETAC books attempt to present the reader with authoritative coverage of the literature, as well as paradigms, methodologies, and controversies; research needs; and new developments specific to the featured topics. The books are generally peer reviewed for SETAC by acknowledged experts.

SETAC publications, which include Technical Issue Papers (TIPs), workshops summaries, newsletter (SETAC Globe), and journals (*Environmental Toxicology and Chemistry* and *Integrated Environmental Assessment and Management*), are useful to environmental scientists in research, research management, chemical manufacturing and regulation, risk assessment, and education, as well as to students considering or preparing for careers in these areas. The publications provide information for keeping abreast of recent developments in familiar subject areas and for rapid introduction to principles and approaches in new subject areas.

SETAC recognizes and thanks the past coordinating editors of SETAC books:

A.S. Green, International Zinc Association
Durham, North Carolina, USA

C.G. Ingersoll, Columbia Environmental Research Center
US Geological Survey, Columbia, Missouri, USA

T.W. La Point, Institute of Applied Sciences
University of North Texas, Denton, Texas, USA

B.T. Walton, US Environmental Protection Agency
Research Triangle Park, North Carolina, USA

C.H. Ward, Department of Environmental Sciences and Engineering
Rice University, Houston, Texas, USA

Contents

List of Figures xiii
List of Tables xv
About the Editors xvii
Workshop Participants xix
Foreword xxi
Acknowledgments xxv

Chapter 1   Introduction 1

Mark Crane, Katie Barrett, and Alistair Boxall
References 3

Chapter 2   Uses and Inputs of Veterinary Medicines in the Environment 7

*Alistair Boxall, Mark Crane, Christian Corsing, Charles Eirkson,
and Alex Tait*
2.1   Introduction 7
2.2   Veterinary Medicine Use 7
     2.2.1 Parasiticides 8
     2.2.2 Antibacterials 8
     2.2.3 Coccidiostats and Antiprotozoals 10
     2.2.4 Antifungals 10
     2.2.5 Aquaculture Medicines 10
     2.2.6 Hormones 10
     2.2.7 Growth Promoters 10
     2.2.8 Other Medicinal Classes 11
2.3   Pathways to the Environment 11
     2.3.1 Emissions during Manufacturing and Formulation 11
     2.3.2 Aquaculture 12
     2.3.3 Agriculture (Livestock Production) 13
     2.3.4 Companion and Domestic Animals 14
     2.3.5 Disposal of Unwanted Drugs 15
2.4   Summary 16
References 17

Chapter 3   Environmental Risk Assessment and Management of
              Veterinary Medicines 21

*Joop de Knecht, Tatiana Boucard, Bryan W. Brooks, Mark Crane,
Charles Eirkson, Sarah Gerould, Jan Koschorreck, Gregor Scheef,
Keith R. Solomon, and Zhixing Yan*

3.1   Introduction 21
3.2   Veterinary Medicines in Regulatory Perspective 23
       3.2.1 Legislation, Scope, and Past Guidelines for Environmental
              Risk Assessment (ERA) of Veterinary Medicines 23
              3.2.1.1 United States 24
              3.2.1.2 European Union 25
              3.2.1.3 Japan 26
              3.2.1.4 Australia 26
              3.2.1.5 Canada 26
       3.2.2 Current Guidelines: VICH and the VICH–EU Technical
              Guidance Document (VICH–EU–TGD) 27
3.3   Refinement of Veterinary Medicinal Product (VMP) Risk Assessments 33
       3.3.1 Metabolism and Degradation 33
       3.3.2 Combination Products 35
       3.3.3 Refinement of Environmental Exposure Predictions 36
       3.3.4 Probabilistic Risk Assessment of Veterinary Medicines 36
              3.3.4.1 Case Study of a Probabilistic Risk Assessment for
                       Dung Fauna 37
3.4   Risk Management 41
       3.4.1 Risk Mitigation Measures within Product Authorization or
              Approval 42
       3.4.2 Risk Assessment and Management beyond Authorization or
              Approval 44
              3.4.2.1 Communication Challenge 44
              3.4.2.2 Incidence Reporting and Pharmacovigilance 47
       3.4.3 Retrospective Risk Assessment 49
       3.4.4 Postmarket Monitoring and Remediation 51
              3.4.4.1 Monitoring Endpoints 51
References 52

Chapter 4   Exposure Assessment of Veterinary Medicines in Aquatic
                Systems 57

*Chris Metcalfe, Alistair Boxall, Kathrin Fenner, Dana Kolpin, Mark Servos,
Eric Silberhorn, and Jane Staveley*

4.1   Introduction 57
4.2   Sources of Veterinary Medicines in the Aquatic Environment 58
       4.2.1 Treatments Used in Agriculture 58
       4.2.2 Treatments Used in Aquaculture 61
4.3   Experimental Studies into the Entry, Fate, and Transport of
       Veterinary Medicines in Aquatic Systems 62
       4.3.1 Aquatic Exposure to Veterinary Medicines Used to Treat
              Livestock 62
              4.3.1.1 Leaching to Groundwater 63
              4.3.1.2 Movement to Surface Water 63

4.3.1.3 Predicting Exposure 65
4.3.1.4 Comparison of Modeled Concentrations with
          Measured Concentrations.. 66
4.3.2 Aquaculture Treatments.. 73
4.3.2.1 Inputs and Fate of Marine Aquaculture Treatments 75
4.3.2.2 Freshwater Aquaculture 76
4.3.2.3 Modeling Exposure from Aquaculture Treatments......... 77
4.4   Conclusions 89
References 91

Chapter 5   Assessing the Aquatic Hazards of Veterinary Medicines............. 97

*Bryan W. Brooks, Gerald T. Ankley, James F. Hobson, James M. Lazorchak,
Roger D. Meyerhoff, and Keith R. Solomon*

5.1   Introduction.. 97
5.2   Protection Goals... 98
5.3   Approaches to Assess Effects of Veterinary Medicines 98
5.3.1 Current Methods of Assessing Aquatic Effects for Risk
        Assessment... 98
5.3.1.1 Lower Tier Approaches 99
5.3.1.2 Higher Tier Testing 99
5.3.1.3 Limitations to Current Approaches............................101
5.3.2 Novel Approaches to Aquatic Effects Assessment 102
5.3.2.1 Use of Chemical Characteristics, Target Organism
          Efficacy Data, Toxicokinetic Data, and Mammalian
          Toxicology Data 102
5.3.2.2 Use of Ecotoxicogenomics in Ecological Effects
          Assessment 108
5.4   Application Factors and Species Sensitivities...110
5.5   Effects of Veterinary Medicines in the Natural Environment...............113
5.5.1 Episodic Exposures...114
5.5.2 Matrix Effects 114
5.5.3 Metabolites and Degradates...115
5.5.4 Mixtures...116
5.5.5 Enantiomer-Specific Hazard...117
5.5.6 Sorption to Sediment 118
5.5.7 Assessing Effects on Communities.......................................119
5.6   Conclusions 121
References 122

Chapter 6   Exposure Assessment of Veterinary Medicines in Terrestrial
Systems.. 129

*Louise Pope, Alistair Boxall, Christian Corsing, Bent Halling-Sørensen,
Alex Tait, and Edward Topp*

6.1   Introduction.. 129
6.2   Absorption and Excretion by Animals.. 130
6.3   Fate during Manure Storage.. 134
6.4   Releases to the Environment... 136
6.5   Factors Affecting Dissipation in the Farm Environment..................... 137
      6.5.1 Dissipation and Transport in Dung Systems............................ 137
      6.5.2 Dissipation and Transport in Soil Systems 138
             6.5.2.1 Biotic Degradation Processes.................................... 138
             6.5.2.2 Abiotic Degradation Processes 140
             6.5.2.3 Sorption to Soil... 141
      6.5.3 Bound Residues.. 141
6.6   Uptake by Plants 143
6.7   Models for Estimating the Concentration of Veterinary Medicine
      in Soil 143
      6.7.1 Intensively Reared Animals.. 144
      6.7.2 Pasture Animals... 148
      6.7.3 PEC Refinement... 148
6.8   Research Needs 149
References.. 149

Chapter 7   Assessing the Effects of Veterinary Medicines on the
               Terrestrial Environment 155

Katie Barrett, Kevin Floate, John Jensen, Joe Robinson, and Neil Tolson
7.1   Introduction... 155
7.2   Considerations Unique to Veterinary Medicines 155
      7.2.1 Routes of Entry 155
      7.2.2 Additional Safety Data Available in the Dossier 156
      7.2.3 Residue Data and Detoxification by the Target Animal
             Species 156
7.3   Protection Goals... 157
7.4   Tiered Testing Strategy 160
7.5   Justification for Existing Testing Methods... 160
7.6   Use of Indicator Species.. 160
7.7   Short-Term and Sublethal Effects Tests 163
7.8   Tier A Testing.. 163
      7.8.1 Physicochemical Properties 163
      7.8.2 Fate... 164
      7.8.3 Microorganisms 164
      7.8.4 Plants.. 165
      7.8.5 Earthworms.. 165
      7.8.6 Collembolans 166
      7.8.7 Dung Fauna.. 166
7.9   Tier B Testing.. 168

7.10 Tier C Testing... 169
        7.10.1 Mesocosm and Field Testing.. 169
        7.10.2 Testing of Additional Species 170
        7.10.3 Monitoring Studies 170
7.11 Calculation of PNEC Concentrations and Use of Assessment
        Factors 171
7.12 Metabolite Testing in Tiers A and B 172
7.13 Secondary Poisoning...173
7.14 Bound Residues...174
7.15 Alternative Endpoints 175
7.16 Modeling Population and Ecosystem Effects (e.g., Bioindicator
        Approaches)..176
7.17 Research Needs 177
References.. 177

Chapter 8 Workshop Conclusions and Recommendations 181

Mark Crane, Katie Barrett, and Alistair Boxall
8.1 Workshop Conclusions...181
8.2 Workshop Recommendations 184

Index... 187

List of Figures

Figure 3.1 VICH phase 1 decision tree..28

Figure 3.2 VICH phase II decision trees. 29

Figure 3.3 Temporal distribution of main seasonal activity of *Aphodius* spp., treatment, and availability of toxically active dung....................39

Figure 3.4 Distribution of effect values in a simple probabilistic model of dung insect toxicity..40

Figure 4.1 Direct and indirect pathways for the release of veterinary medicines into the aquatic environment..58

Figure 4.2 Comparison of predicted pore water concentrations with measured maximum concentrations in leachate, groundwater, drainflow, and runoff water for 8 veterinary medicines for which measured concentrations were available in field and semifield studies....................67

Figure 4.3 Comparison of predicted surface water concentrations with measured concentrations for surface water for 9 veterinary medicines for which measured concentrations were available in field studies.................70

Figure 4.4 Comparison of VetCalc predictions of environmental concentration in soil (PEC$_{soil}$) under 12 scenarios with data on measured soil concentrations (MEC$_{soil}$). 73

Figure 4.5 Comparison of VetCalc predictions of environmental concentration in surface water (PEC$_{surface water}$) under 12 scenarios with data on measured surface water concentrations (MEC$_{surface water}$). 73

Figure 4.6 Schematic of a typical flow-through aquaculture facility showing the basic and optional components of the system. 74

Figure 5.1 Screening assessment approach to target aquatic effects testing with fish from water exposure. 105

Figure 5.2 Species sensitivity distributions for aquatic organisms exposed to an antibiotic in water. 112

Figure 6.1 Excretion profiles of ivermectin following 3 different application methods...132

Figure 6.2 The percentage of the applied dose excreted in the dung (in black) and urine (in gray), as parent molecule and/or metabolites....................133

Figure 6.3 Measured and predicted environmental concentrations (MEC and PEC) for a range of veterinary medicines. 147

Figure 7.1 Abiotic and biotic factors that affect the degradation of cattle dung pats on pasture..168

Figure 7.2 Screening schemes for testing metabolites and soil degradates. 173

List of Tables

Table 2.1   Major groups of veterinary medicines 9

Table 3.1   Overview of the regulatory situation for environmental risk assessment of veterinary medicines 24

Table 3.2   International Cooperation on Harmonization of Technical Requirements for Registration of Veterinary Products (VICH) tier A fate and effects studies to be included...30

Table 3.3   International Cooperation on Harmonization of Technical Requirements for Registration of Veterinary Products (VICH) tier B effects studies 31

Table 3.4   Parameters for estimating parasiticide impacts on dung insect populations.. 40

Table 3.5   Criteria for classifying known or predicted effects of veterinary medicines in the ecosystem 42

Table 4.1   Major sources of veterinary medicines and the activities leading to exposure in aquatic environments 60

Table 4.2   Field scale and column studies reported in the literature on the fate and transport of veterinary medicines...68

Table 4.3   Input data on chemical and physical parameters of veterinary medicines used in modeling exercises...71

Table 5.1   Tier B tests proposed by the International Cooperation on Harmonization of Technical Requirements for Registration of Veterinary Products (VICH)..100

Table 5.2   Example scenarios for veterinary medicines where aquatic hazards might or might not be found by current regulatory toxicity-testing approaches with standard endpoints... 101

Table 5.3   Examples of how the results from mammalian tests can be used to target environmental effects testing 103

Table 5.4   Physicochemical characteristics of emamectin benzoate 106

Table 5.5   Predicted no-effect concentrations (PNECs) for aquatic organisms exposed to an antibiotic 111

Table 5.6   Typical types and characteristics of cosms 120

Table 6.1   General trend for the degree of metabolism of major therapeutic classes of veterinary medicines...130

Table 6.2   Parasiticide formulations available in the United Kingdom 131

Table 6.3   Commonly employed practices for manure storage and handling..............135

Table 6.4   Characteristics of manure type or application of best management practices (BMP) that can influence the persistence of veterinary medicines in soil 136

Table 6.5   Mobility and persistence of veterinary medicines, classification of persistence, and mobility...139

Table 6.6   Comparison of predicted environmental concentration in soil (PEC_{soil}) values using different calculation methods obtained for a hypothetical veterinary medicine dosed at 10 mg kg^{-1} 147

Table 7.1 Changing emphasis of protection goals across a gradient of land use: illustrated with four categories 159

Table 7.2 Generic study designs for tiers A to C 161

About the Editors

Mark Crane, PhD, is a director of Watts & Crane Associates (www.wca-environment.com). He has a first degree in ecology and a PhD in ecotoxicology and has worked on the effects of chemicals on wildlife for more than 19 years, in both consulting and academia. Crane has edited 3 books and published more than 100 papers on environmental toxicology and risk assessment, including research on endocrine-disrupting pharmaceuticals. Recently, Crane's work in human and veterinary medicines has included advice to industry clients on preparation of environmental risk assessments, statistical analysis of monitoring data, and reviews for the Environment Agency of England and Wales on chronic ecotoxicity test methods for medicines, and for the UK Department for Environment, Food and Rural Affairs on the occurrence of medicines in surface waters. Crane served for 4 years as the environmental expert on the UK Veterinary Products Committee.

Alistair Boxall, PhD, currently leads the joint University of York–Central Science Laboratory EcoChemistry Team (www.csl.gov.uk). He specializes in environmental chemistry and has research interests in the fate, behavior, and effects of pesticides, biocides, veterinary medicines, industrial chemicals, and nanomaterials in the environment. Boxall has previously worked at the Plymouth Marine Laboratory, the University of Sheffield, Liverpool John Moores University, and the Water Research Centre, and, more recently, Cranfield University, where he was joint head of the Cranfield Centre for EcoChemistry. He is cur-

rently or has been a member of professional bodies, including the UK Veterinary Products Committee, the European Food Safety Authority (EFSA) ad hoc committee on risk assessment of feed additives, the Royal Society of Chemistry (RSC) expert group on water, and the UK government working group on exposure assessment of nanomaterials. From 1999 to 2003, he coordinated an EU project on environmental risk assessment of veterinary medicines.

Katie Barrett, PhD, worked for 18 years for AgEvo (formerly Hoechst and Schering); working initially in the environmental metabolism department, she was also responsible for setting up the ecotoxicology group. She joined Huntingdon Life Sciences (www. huntingdon.com) in June 1995 as head of the Ecotoxicology Department and is now program director for agrochemical and veterinary programs. She is also actively involved in liaising on behalf of clients with regulatory authorities and preparing risk assessments for both veterinary and agrochemical products. Barrett has served on a number of working groups for the Organization for Economic Cooperation and Development (OECD) and SETAC, developing guidance documents and guidelines for novel test species, including sediment organisms, dung fauna, and beneficial insects. She is currently a member of the UK OECD shadow group, commenting on new draft guidelines.

Workshop Participants

Gerald T. Ankley
US Environmental Protection Agency
Duluth, Minnesota, USA

Katie Barrett
Huntingdon Life Sciences
Huntingdon, Cambridgeshire, UK

Tatiana Boucard
Environment Agency of England and
    Wales
Wallingford, Oxfordshire, UK

Alistair Boxall
Central Science Laboratory
University of York
York, UK

Bryan W. Brooks
Baylor University
Waco, Texas, US

Christian Corsing
Bayer Healthcare
Monheim, Germany

Mark Crane
Watts & Crane Associates
Faringdon, Oxfordshire, UK

Joop de Knecht
Dutch National Institute for Public
    Health and the Environment (RIVM)
Bilthoven, The Netherlands

Charles Eirkson
US Food and Drug Administration
Washington, DC, USA

Kathrin Fenner
Eawag Dübendorf/ETH
Zürich, Switzerland

Kevin Floate
Agriculture and Agri-Food Canada
Lethbridge, Alberta, Canada

Sarah Gerould
US Geological Survey
Reston, Virginia, USA

Bent Halling-Sørensen
University of Copenhagen
Copenhagen, Denmark

James F. Hobson
MorningStar Consulting
Germantown, Maryland, USA

John Jensen
National Environmental Research
    Institute
University of Aarhus
Silkeborg, Denmark

Dana Kolpin
US Geological Survey
Iowa City, Iowa, USA

Jan Koschorreck
Umweltbundesamt (UBA; Federal
    Environment Agency)
Dessau, Germany

James M. Lazorchak
US Environmental Protection Agency
Cincinnati, Ohio, USA

Chris Metcalfe
Trent University
Peterborough, Ontario, Canada

Roger D. Meyerhoff
Eli Lilly & Company
Indianapolis, Indiana, USA

Louise Pope
University of York
York, UK

Joe Robinson
Pfizer Inc.
Kalamazoo, Missouri, USA

Gregor Scheef
Intervet Innovation GmbH
Schwabenheim, Germany

Mark Servos
University of Waterloo
Waterloo, Ontario, Canada

Eric Silberhorn
US Food and Drug Administration
Washington, DC, USA

Keith R. Solomon
University of Guelph
Guelph, Ontario, Canada

Jane Staveley
Arcadis Consulting
Durham, North Carolina, USA

Alex Tait
Veterinary Medicines Directorate
New Haw, Surrey, UK

Neil Tolson
Health Canada
Ottawa, Ontario, Canada

Edward Topp
Agriculture and Agri-Food Canada
London, Ontario, Canada

Zhixing Yan
Merial Limited
North Brunswick, New Jersey, USA

Foreword

The workshop from which this book resulted, Veterinary Medicines in the Environment, held in Pensacola, Florida, 12–16 February 2006, was part of the successful "Pellston Workshop Series." Since 1977, Pellston Workshops have brought scientists together to evaluate current and prospective environmental issues. Each workshop has focused on a relevant environmental topic, and the proceedings of each have been published as peer-reviewed or informal reports. These documents have been widely distributed and are valued by environmental scientists, engineers, regulators, and managers for their technical basis and their comprehensive, state-of-the-science reviews. The other workshops in the Pellston series are as follows:

- Estimating the Hazard of Chemical Substances to Aquatic Life. Pellston, Michigan, 13–17 Jun 1977. Published by the American Society for Testing and Materials, STP 657, 1978.
- Analyzing the Hazard Evaluation Process. Waterville Valley, New Hampshire, 14–18 Aug 1978. Published by The American Fisheries Society, 1979.
- Biotransformation and Fate of Chemicals in the Aquatic Environment. Pellston, Michigan, 14–18 Aug 1979. Published by The American Society of Microbiology, 1980.
- Modeling the Fate of Chemicals in the Aquatic Environment. Pellston, Michigan, 16–21 Aug 1981. Published by Ann Arbor Science, 1982.
- Environmental Hazard Assessment of Effluents. Cody, Wyoming, 23–27 Aug 1982. Published as a SETAC Special Publication by Pergamon Press, 1985.
- Fate and Effects of Sediment-Bound in Aquatic Systems. Florissant, Colorado, 11–18 Aug 1984. Published as a SETAC Special Publication by Pergamon Press, 1987.
- Research Priorities in Environmental Risk Assessment. Held in Breckenridge, Colorado, 16–21 Aug 1987. Published by SETAC, 1987.
- Biomarkers: Biochemical, Physiological, and Histological Markers of Anthropogenic Stress. Keystone, Colorado, 23–28 Jul 1989. Published as a SETAC Special Publication by Lewis Publishers, 1992.
- Population Ecology and Wildlife Toxicology of Agricultural Pesticide Use: A Modeling Initiative for Avian Species. Kiawah Island, South Carolina, 22–27 Jul 1990. Published as a SETAC Special Publication by Lewis Publishers, 1994.

- A Technical Framework for [Product] Life-Cycle Assessments. Smuggler's Notch, Vermont, 18–23 Aug 1990. Published by SETAC, Jan 1991; 2nd printing Sep 1991; 3rd printing Mar 1994.
- Aquatic Microcosms for Ecological Assessment of Pesticides. Wintergreen, Virginia, 7–11 Oct 1991. Published by SETAC, 1992.
- A Conceptual Framework for Life-Cycle Assessment Impact Assessment. Sandestin, Florida, 1–6 Feb 1992. Published by SETAC, 1993.
- A Mechanistic Understanding of Bioavailability: Physical–Chemical Interactions. Pellston, Michigan, 17–22 Aug 1992. Published as a SETAC Special Publication by Lewis Publishers, 1994.
- Life-Cycle Assessment Data Quality Workshop. Wintergreen, Virginia, 4–9 Oct 1992. Published by SETAC, 1994.
- Avian Radio Telemetry in Support of Pesticide Field Studies. Pacific Grove, California, 5–8 Jan 1993. Published by SETAC, 1998.
- Sustainability-Based Environmental Management. Pellston, Michigan, 25–31 Aug 1993. Co-sponsored by the Ecological Society of America. Published by SETAC, 1998.
- Ecotoxicological Risk Assessment for Chlorinated Organic Chemicals. Alliston, Ontario, Canada, 25–29 Jul 1994. Published by SETAC, 1998.
- Application of Life-Cycle Assessment to Public Policy. Wintergreen, Virginia, 14–19 Aug 1994. Published by SETAC, 1997.
- Ecological Risk Assessment Decision Support System. Pellston, Michigan, 23–28 Aug 1994. Published by SETAC, 1998.
- Avian Toxicity Testing. Pensacola, Florida, 4–7 Dec 1994. Co-sponsored by Organisation for Economic Co-operation and Development. Published by OECD, 1996.
- Chemical Ranking and Scoring (CRS): Guidelines for Developing and Implementing Tools for Relative Chemical Assessments. Sandestin, Florida, 12–16 Feb 1995. Published by SETAC, 1997.
- Ecological Risk Assessment of Contaminated Sediments. Pacific Grove, California, 23–28 Apr 1995. Published by SETAC, 1997.
- Ecotoxicology and Risk Assessment for Wetlands. Fairmont, Montana, 30 Jul–3 Aug 1995. Published by SETAC, 1999.
- Uncertainty in Ecological Risk Assessment. Pellston, Michigan, 23–28 Aug 1995. Published by SETAC, 1998.
- Whole-Effluent Toxicity Testing: An Evaluation of Methods and Prediction of Receiving System Impacts. Pellston, Michigan, 16–21 Sep 1995. Published by SETAC, 1996.
- Reproductive and Developmental Effects of Contaminants in Oviparous Vertebrates. Fairmont, Montana, 13–18 Jul 1997. Published by SETAC, 1999.
- Multiple Stressors in Ecological Risk Assessment. Pellston, Michigan, 13–18 Sep 1997. Published by SETAC, 1999.

- Re-evaluation of the State of the Science for Water Quality Criteria Development. Fairmont, Montana, 25–30 Jun 1998. Published by SETAC, 2003.
- Criteria for Persistence and Long-Range Transport of Chemicals in the Environment. Fairmont Hot Springs, British Columbia, Canada, 14–19 Jul 1998. Published by SETAC. 2000.
- Assessing Contaminated Soils: From Soil-Contaminant Interactions to Ecosystem Management. Pellston, Michigan, 23–27 Sep 1998. Published by SETAC, 2003.
- Endocrine Disruption in Invertebrates: Endocrinology, Testing, and Assessment (EDIETA). Amsterdam, The Netherlands, 12–15 Dec 1998. Published by SETAC, 1999.
- Assessing the Effects of Complex Stressors in Ecosystems. Pellston, Michigan, 11–16 Sep 1999. Published by SETAC, 2001.
- Environmental–Human Health Interconnections. Snowbird, Utah, 10–15 Jun 2000. Published by SETAC, 2002.
- Ecological Assessment of Aquatic Resources: Application, Implementation, and Communication. Pellston, Michigan, 16–21 Sep 2000. Published by SETAC, 2004.
- The Global Decline of Amphibian Populations: An Integrated Analysis of Multiple Stressors Effects. Wingspread, Racine, Wisconsin, 18–23 Aug 2001. Published by SETAC, 2003.
- Methods of Uncertainty Analysis for Pesticide Risks. Pensacola, Florida, 24 Feb–1 Mar 2002.
- The Role of Dietary Exposure in the Evaluation of Risk of Metals to Aquatic Organisms. Fairmont Hot Springs, British Columbia, Canada, 27 Jul–1 Aug 2002. Published by SETAC, 2005.
- Use of Sediment Quality Guidelines (SQGs) and Related Tools for the Assessment of Contaminated Sediments. Fairmont Hot Springs, Montana, 17–22 Aug 2002. Published by SETAC, 2005.
- Science for Assessment of the Impacts of Human Pharmaceuticals on Aquatic Ecosystem. Held in Snowbird, Utah, 3–8 Jun 2003. Published by SETAC, 2005.
- Population-Level Ecological Risk Assessment. Held in Roskilde, Denmark, 23-27 Aug 2003. Published by SETAC and CRC Press, 2007.
- Valuation of Ecological Resources: Integration of Ecological Risk Assessment and Socio-Economics to Support Environmental Decisions. Pensacola, Florida, 4–9 Oct 2003. Published by SETAC and CRC Press, 2007.
- Emerging Molecular and Computational Approaches for Cross-Species Extrapolations. Portland, Oregon, 18–22 Jul 2004. Published by SETAC and CRC Press, 2006.

- Tissue Residue Approach for Toxicity Assessment: Invertebrates and Fish. Leavenworth, Washington, 7–10 Jun 2007. To be published by SETAC and CRC Press, 2008.
- Science-Based Guidance and Framework for the Evaluation and Identification of PBTs and POPs. Pensacola Beach, Florida, 27 Jan–1 Feb 2008.

Acknowledgments

This book presents the proceedings of a technical workshop convened by the Society of Environmental Toxicology and Chemistry (SETAC) in Pensacola, Florida, USA, in February 2006. The 31 scientists involved in this workshop represented 8 countries and offered expertise in ecology, ecotoxicology, environmental chemistry, environmental regulation, and risk assessment. Their goals were to examine the current state of science in evaluating the potential risks of veterinary medicines to aquatic and terrestrial ecosystems, and to make recommendations on how this science can be used to inform regulations.

The workshop was made possible by the generous support of many organizations, including the following:

- Department for Environment, Food and Rural Affairs (United Kingdom)
- Elanco
- Environment Agency of England and Wales
- Health Canada
- Intervet
- Pfizer
- UBA (Germany)
- US Environmental Protection Agency
- US Geological Survey

We are also grateful to Professor Peter Matthiessen for expert and helpful peer review of the final draft chapters in this book.

1 Introduction

Mark Crane, Katie Barrett, and Alistair Boxall

Potential risks associated with releases of medicines into the environment have become an increasingly important issue for environmental regulators (Jørgensen and Halling-Sørensen 2000; Stuer-Lauridsen et al. 2000; Kümmerer 2004). This concern has been driven by widespread detection of human and veterinary medicines in environmental samples as a result of improved analytical capabilities and the commissioning of focused field surveys (Daughton 2001; Focazio et al. 2004; Webb 2004). Surface water-sampling programs in Europe (e.g., Buser et al. 1998; Ternes 1998; Calamari et al. 2003; Thomas and Hilton 2003; Alder et al. 2004; Ashton et al. 2004; Zuccato et al. 2004), North America (e.g., Kolpin et al. 2002; Metcalfe et al. 2003, 2004; Anderson et al. 2004; Focazio et al. 2004), and elsewhere (Heberer 2002) have shown the presence of many different classes of medicines, some of which are known to be environmentally persistent (Zuccato et al. 2004). Although some of these medicines are unlikely to be a risk to the environment because of low concentrations combined with low toxicity, others may pose considerable risks.

A Society of Environmental Toxicology and Chemistry (SETAC) workshop was held in Snowbird, Utah, in 2003 to assess the state of the art in evaluating the impacts of human medicines on nontarget species in aquatic ecosystems (Williams 2006). Medicines used in both veterinary and human medicine have been a focus of regulatory attention, but environmental exposure scenarios differ substantially between the two. Exposure of wildlife to human medicines is most likely to occur from sewage treatment works discharges into the aquatic environment (Focazio et al. 2004), and this exposure may therefore be at continuous, low concentrations (Daughton and Ternes 1999; Breton and Boxall 2003). In contrast, exposure to veterinary medicines is likely to be via a wider range of point and diffuse sources, with environmental pathways from treated animals into both aquatic and terrestrial habitats (Boxall et al. 2004). Guidance based on standard risk assessment approaches is available on how to assess the environmental effects of veterinary medicines, as discussed in Chapter 3 of this book. However, one aspect of medicines that distinguishes them from many other classes of chemicals is that regulatory submissions from manufacturers also usually contain large amounts of additional information on modes and mechanisms of action and the adsorption, distribution, metabolism, and elimination (ADME) of the medicine in the body of target animals. These data may be of substantial use in identifying potentially sensitive nontarget species and for extrapolating from target species to effects on these nontarget species (e.g., Huggett et al. 2002, 2003, 2004).

This book reports on the findings from a SETAC Workshop held in Pensacola, Florida, in February 2006, which followed on from and complements the earlier workshop on human medicines in the environment (Williams 2006). The SETAC Workshop on Veterinary Medicines in the Environment assessed the current state of science in evaluating the potential risks of veterinary medicines to aquatic and terrestrial ecosystems, particularly from those medicines used to treat food-producing species. The workshop followed the standard SETAC format, bringing together more than 30 experts from 8 countries with expertise in risk assessment, environmental toxicology and chemistry, and environmental policy and regulation. Participants were drawn from academic, government, and business sectors.

The main aim of the workshop was to examine the current state of science and provide recommendations in 5 areas:

1) Risk assessment, management, and communication for veterinary medicines in the environment
2) Exposure assessment of veterinary medicines in the terrestrial environment
3) Effects assessment of veterinary medicines in the terrestrial environment
4) Exposure assessment of veterinary medicines in the aquatic environment
5) Effects assessment of veterinary medicines in the aquatic environment

The specific objectives of the meeting were as follows:

1) To review the major classes of veterinary drugs (including coccidiostats) and determine whether they are adequately covered by current regulatory guidance
2) To identify environmental fate and effects study types recommended under existing regulatory guidance and recommend any appropriate changes or additions
3) To assess whether information from other parts of a regulatory submission can be used to assess environmental effects or bioaccumulation potential with read-across, quantitative structure-activity relationship (QSAR), or other modeling approaches
4) To recommend appropriate tests, data, or risk mitigation measures that should be considered if an assessment still indicates a risk at the end of current risk assessment procedures
5) To advise on how to assess cumulative impacts (e.g., multiple sites or products) and possible mixture toxicity effects
6) To advise on when, how, and what risk management and communications should be utilized for veterinary medicinal products within the current regulatory frameworks
7) In the light of existing approaches to determining risks from veterinary medicinal products, as well as changes recommended in this workshop, to identify future areas for research to improve our understanding of the potential for veterinary medicines to impact the environment

The focus of the meeting was on those products, sources, pathways, and receptors likely to present the greatest potential for environmental effects. Uses of veterinary medicines on companion animals and treatment of individual food-producing animals were not, therefore, considered in detail. Because of our focus on possible environmental effects, we also excluded consideration of human health issues, such as the potential for veterinary antimicrobial products to induce resistance to antimicrobials used in human medicine.

This book begins with an overview of veterinary medicine use and characteristics, and consideration of current regulatory drivers, their protection goals, and the associated risk assessment and management frameworks. We then consider the pathways along which veterinary medicines may travel from target animals and into the wider terrestrial and aquatic environment and what influences the fate and behavior of medicines along these pathways. The potential effects of veterinary medicines on organisms in the environment are then considered by reviewing biological tools and techniques that provide information on toxicity at different levels of biological complexity. Finally, we end with a list of overall conclusions from the workshop and recommendations for further research and development to advance this scientific field.

We could not have produced this book without the assistance of an excellent steering committee, the SETAC staff, and the full engagement of the workshop participants, all of whom are coauthors of the remaining chapters. We thank them for their superb contributions to what we believe is an authoritative and integrated text for graduate students and professionals in the field of environmental science with an interest in veterinary medicines in the environment.

REFERENCES

Alder AC, McArdell CS, Golet EM, Kohler HPE, Molnar E, Anh Pham Thi N, Siegrist H, Suter MJF, Giger W. 2004. Environmental exposure of antibiotics in wastewaters, sewage sludges and surface waters in Switzerland. In: Kümmerer K, editor. Pharmaceuticals in the environment: sources, fate, effects and risks. 2nd ed. Berlin (Germany): Springer, p 55–66.

Anderson PD, D'Aco VJ, Shanahan P, Chapra SC, Buzby ME, Cunningham VL, DuPlessie BM, Hayes EP, Mastrocco F, Parke NJ, Rader JC, Samuelian JH, Schwab BW. 2004. Screening analysis of human pharmaceutical compounds in US surface waters. Environ Sci Technol 38:838–849.

Ashton D, Hilton M, Thomas KV. 2004. Investigating the environmental transport of human pharmaceuticals to streams in the United Kingdom. Sci Total Environ 333:167–184.

Boxall ABA, Fogg LA, Blackwell PA, Kay P, Pemberton EJ, Croxford A. 2004. Veterinary medicines in the environment. Rev Environ Contam Toxicol 180:1–91.

Breton R, Boxall A. 2003. Pharmaceuticals and personal care products in the environment: regulatory drivers and research needs. QSAR Comb Sci 22:399–409.

Buser HR, Müller MD, Theobald N. 1998. Occurrence of the pharmaceutical drug clofibric acid and the herbicide mecoprop in various Swiss lakes and in the North Sea. Environ Sci Technol 32:188–192.

Calamari D, Zuccato E, Castiglioni S, Bagnati R, Fanelli R. 2003. Strategic survey of therapeutic drugs in the Rivers Po and Lambro in Northern Italy. Environ Sci Technol 37:1241–1248.

Daughton CG. 2001. Pharmaceuticals in the environment: overarching issues and overview. In: Daughton CG, Jones-Lepp T, editors. Pharmaceuticals and personal care products in the environment: scientific and regulatory issues. Symposium Series 791. Washington (DC): American Chemical Society, p 2–38.

Daughton CG, Ternes TA. 1999. Pharmaceuticals and personal care products in the environment: agents of subtle change? Environ Health Perspect 107:907–938.

Focazio MJ, Kolpin DW, Furlong ET. 2004. Occurrence of human pharmaceuticals in water resources of the United States: a review. In: Kümmerer K, editor. Pharmaceuticals in the environment: sources, fate, effects and risks. 2nd ed. Berlin (Germany): Springer, p 91–105.

Heberer T. 2002. Occurrence, fate, and removal of pharmaceutical residues in the aquatic environment: a review of recent research data. Toxicol Lett 131:5–17.

Huggett DB, Brooks BW, Peterson B, Foran CM, Schlenk D. 2002. Toxicity of selected beta adrenergic receptor blocking pharmaceuticals (β-blockers) on aquatic organisms. Arch Environ Contam Toxicol 43:229–235.

Huggett DB, Cook JC, Ericson JF, Williams RT. 2003. Theoretical model for prioritizing potential impacts of human pharmaceuticals to fish. Human Ecol Risk Assess 9:1789–1799.

Huggett DB, Ericson JF, Cook JC, Williams RT. 2004. Plasma concentrations of human pharmaceuticals as predictors of pharmacological responses in fish. In: Kümmerer K, editor. Pharmaceuticals in the environment: sources, fate, effects and risks. 2nd ed. Berlin (Germany): Springer, p 373–386.

Jørgensen SE, Halling-Sørensen B. 2000. Drugs in the environment. Chemosphere 40:691–699.

Kolpin DW, Furlong ET, Meyer MT, Thurman EM, Zaugg SD, Barber LB, Buxton HT. 2002. Pharmaceuticals, hormones, and other organic wastewater contaminants in US streams 1999–2000: a national reconnaissance. Environ Sci Technol 36:1202–1211.

Kümmerer K. 2004. Pharmaceuticals in the environment: sources, fate, effects and risks. 2nd ed. Berlin (Germany): Springer.

Metcalfe C, Miao XS, Hua W, Letcher R, Servos M. 2004. Pharmaceuticals in the Canadian environment. In: Kümmerer K, editor. Pharmaceuticals in the environment: sources, fate, effects and risks. 2nd ed. Berlin (Germany): Springer, p 67–90.

Metcalfe CD, Miao XS, Koenig BG, Struger J. 2003. Distribution of acidic and neutral drugs in surface waters near sewage treatment plants in the lower Great Lakes, Canada. Environ Toxicol Chem 22:2881–2889.

Stuer-Lauridsen F, Birkved M, Hansen LP, Holten Lützhøft HC, Halling-Sørensen B. 2000. Environmental risk assessment of human pharmaceuticals in Denmark after normal use. Chemosphere 40:783–793.

Ternes TA. 1998. Occurrence of drugs in German sewage treatment plants and rivers. Water Res 12:3245–3260.

Thomas KV, Hilton M. 2003. Targeted monitoring programme for pharmaceuticals in the aquatic environment. R&D Technical Report P6-012/6. Bristol (UK): Environment Agency.

Webb SF. 2004. A data-based perspective on the environmental risk assessment of human pharmaceuticals II: aquatic risk characterisation. In: Kümmerer K, editor. Pharmaceuticals in the environment: sources, fate, effects and risks. 2nd ed. Berlin (Germany): Springer, p 345–361.

Williams R. 2006. Human pharmaceuticals in the environment. Pensacola (FL): SETAC Press.

Zuccato E, Castiglioni S, Fanelli R, Bagnati R, Calamari D. 2004. Pharmaceuticals in the environment: changes in the presence and concentrations of pharmaceuticals for human use in Italy. In: Kümmerer K, editor. Pharmaceuticals in the environment: sources, fate, effects and risks. 2nd ed. Berlin (Germany): Springer, p 45–53.

2 Uses and Inputs of Veterinary Medicines in the Environment

Alistair Boxall, Mark Crane, Christian Corsing, Charles Eirkson, and Alex Tait

2.1 INTRODUCTION

Veterinary medicines are widely used to treat disease and protect the health of animals. Dietary-enhancing feed additives (growth promoters) are also incorporated into the feed of animals reared for food in order to improve their growth rates. Release of veterinary medicines to the environment occurs directly, for example, from the use of medicines in fish farms. It also occurs indirectly, via the application of animal manure (containing excreted products) to land or via direct excretion of residues onto pasture (Jørgensen and Halling-Sørensen 2000; Boxall et al. 2004).

Over the past 10 years, the scientific community has become increasingly interested in the impacts of veterinary medicines on the environment, and there have been significant developments in the regulatory requirements for the environmental assessment of veterinary products. A number of groups of veterinary medicines, primarily sheep dip chemicals (Environment Agency 1997), fish farm medicines (Jacobsen and Berglind 1988; Davies et al. 1998), and anthelmintics (Wall and Strong 1987; Ridsdill-Smith 1988; McCracken 1993; Strong 1993; McKellar 1997), have been well studied.

This chapter considers publicly available data on the use and inputs to the environment of veterinary medicines and provides an overall context for subsequent chapters in this book.

2.2 VETERINARY MEDICINE USE

Data on amounts used and sales of veterinary medicines are available from several sources, including survey data obtained from Intercontinental Medical Statistics (IMS) Health, the UK Veterinary Medicines Directorate (VMD) data on the sales of antimicrobial substances and sheep dip chemicals in the United Kingdom and data in the published literature (e.g., Sarmah et al. 2006; Kools et al. 2008). It is not, however, possible to obtain a complete data set for usage of all veterinary medicines. However, taken together these data sets are likely to reflect the general picture of usage of veterinary medicines in Europe, North America, and elsewhere. Major active substances used are shown in Table 2.1.

2.2.1 PARASITICIDES

Ectoparasiticides are used to control external parasites in livestock. Endoparasiticides are used to control internal parasites, and endectocides are used to treat both internal and external parasites. Ectoparasiticides, endoparasiticides, and endectocides are used to treat parasites in a wide range of animals. If uncontrolled, ectoparasites (mites, blowflies, lice, ticks, headflies, and keds) can severely affect the welfare of farm animals. Several product types are available, and a range of active substances is approved for use (Table 2.1). Kools et al. (2008) estimated that approximately 194 tons of parasiticides are used in Europe in 1 year, but data on usage of individual active substances are limited. The available data on the usage of ectoparasiticides on sheep (Liddel 2000; Pepper and Carter 2000) indicate that the organophosphate compound diazinon is the most widely used active ingredient, followed by the synthetic pyrethroids such as cypermethrin. Data from the United Kingdom indicate that in cattle, the most widely used parasiticide is ivermectin, followed by oxfendazole, eprinomectin, doramectin, and fenbendazole, with morantel, moxidectin, and permethrin used in much lower amounts (Boxall et al. 2007).

2.2.2 ANTIBACTERIALS

Antibacterials are used in the treatment and prevention of bacterial diseases (Gustafson and Bowen 1997). Although their veterinary use follows similar principles to those used in human medicines, there are some differences. The most significant is that livestock and poultry are raised in large numbers, and it is therefore necessary to treat the entire flock or herd at risk. An extensive review of antibacterial use across the world is provided in Sarmah et al. (2006). In the United States, it is estimated that 16000 tons of antimicrobial compounds are used annually. These include ionophores, sulfonamides, tetracyclines, fluoroquinolones, β-lactams, and aminoglycosides. In the European Union, approximately 5400 tons of antibiotics are used per year (Kools et al. 2008). The type of antibacterial used depends on the EU member state. For example, in the United Kingdom, the Netherlands, and France, the tetracyclines are the biggest usage class, whereas in Sweden, Finland, and Denmark, the β-lactams and cephalosporins comprise the largest usage class (Kools et al. 2008). In New Zealand, 93 tons of antibiotics are used per year, the majority of which are ionophores (Sarmah et al. 2006). In Kenya, around 15 tons of antibiotics are used per year, the majority of which are tetracyclines and potentiated sulfonamides (i.e., the products contain a mixture of a sulfonamide and trimethoprim; Sarmah et al. 2006).

TABLE 2.1

Major groups of veterinary medicines

Group	Chemical class	Major active ingredients
Antibacterials	Tetracyclines	Oxytetracycline, chlortetracycline, tetracycline
	Sulphonamides	Sulfadiazine, sulfamethazine, sulfathiazole
	β-lactams	Amoxicillin, ampicillin, penicillin G, benzylpenicillin
	Aminoglycosides	Dihydrostreptomycin, neomycin, apramycin
	Macrolides	Tylosin, spiramycin, erythromycin, lincomycin
	Fluoroquinolones	Enrofloxacin
	2,4-diaminopyrimidines	Trimethoprim
	Pleuromutilins	Tiamulin
Parasiticides	Macrolide endectins	Ivermectin, doramectin, eprinomectin
	Pyrethroids	Cypermethrin, deltamethrin
	Organophosphates	Diazinon
	Pyrimidines	Pyrantel, morantel
	Benzimidazoles	Triclabendazole, fenbendazole
	Others	Levamisole
Hormones		Altrenogest, progesterone, medroxyprogesterone, methyltestosterone, estradiol benzoate
Antifungals	Biguanide/gluconate	Chlorhexidine
	Azole	Miconazole
	Other	Griseofulvin
Coccidiostats/ antiprotozoals		Amprolium, clopidol, lasalocid, maduramicin, narasin, nicarbazin, robenidine, toltrazuril, diclazuril
Growth promoters		Monensin, salinomycin, flavophospholipol
Aquaculture treatments		Oxytetracycline, amoxicillin, florfenicol, emamectin benzoate, cypermethrin, teflubenzuron, hydrogen peroxide
Anaesthetics		Isoflurane, halothane, procaine, lido/lignocaine
Euthanasia products		Pentobarbitone
Analgesics		Metamyzole
Tranquilizers		Phenobarbitone
Nonsteroidal anti-inflammatory drugs (NSAIDs)		Phenylbutazone, caprofen
Enteric bloat preps		Dimethicone, poloxalene

2.2.3   COCCIDIOSTATS AND ANTIPROTOZOALS

Coccidiostats and antiprotozoals are often incorporated into feedstuffs for medicinal purposes. This includes prophylactic use for the prevention of diseases such as coccidiosis and swine dysentery and therapeutic use for the treatment of diseases. Apart from 1 individual substance (dimetridazole), usage data are largely unavailable (Boxall et al. 2004). However, the following compounds are considered to be potential major usage compounds within the therapeutic group: amprolium, clopidol, lasalocid acid, maduramicin, narasin, nicarbazin, and robenidine hydrochloride. Major usage protozoal compounds include toltrazuril, decoquinate, and diclazuril.

2.2.4   ANTIFUNGALS

Antifungal agents are used topically and orally to treat fungal and yeast infections. The most common uses include treatment of ringworm and yeast infections. The publicly available data indicate that the major active substances used are chlorhexidine, miconazole, and griseofulvin.

2.2.5   AQUACULTURE MEDICINES

A range of substances are used in aquaculture to treat mainly sea lice infestations and furunculosis. The medicines may be applied by injection, in feed, or via cage treatments. A range of substances are used, including oxytetracycline, oxolinic acid, amoxicillin, co-trimazine, florfenicol, sarafloxacin, emamectin benzoate, cypermethrin, deltamethrin, teflubenzuron, azamethiphos, and hydrogen peroxide.

2.2.6   HORMONES

Although they are currently banned as growth promoters in the European Union, hormones have other restricted uses, including induction of ovulatory estrus, suppression of estrus, systemic progesterone therapy, and treatment of hypersexuality. It has been estimated that in the European Union, the amount of hormones used in animal treatment is around 4.5 tons per year (Kools et al. in press). The major active substances used are altrenogest and progesterone.

2.2.7   GROWTH PROMOTERS

Growth promoters (also called "digestive enhancers") are mainly antibiotic compounds added to animal feedstuffs to improve the efficiency of food digestion. From 1993 to 1998, sales of antimicrobial growth promoters remained largely static. However, in 1999, sales fell by 69%. This decrease is considered to be due to a ban by the European Union in mid-1999 of those growth promoters suspected to confer cross-resistance to antimicrobials in human medicine (VMD

2001), although use of growth promoters continues in other regions of the world. Usage data on individual antimicrobial compounds used as growth promoters are limited. Compounds identified as potentially major usage growth promoters include monensin, flavophospolipol, and salinomycin sodium.

2.2.8   OTHER MEDICINAL CLASSES

Several other therapeutic groups are used as veterinary medicines in significant quantities, including anesthetics, euthanasia products, analgesics, tranquilizers, nonsteroidal anti-inflammatory drugs (NSAIDs), and enteric preparations.

In addition to the above, the following "other" therapeutic groups have also been identified as potentially important: antiseptics, steroids, diuretics, cardiovascular and respiratory treatments, locomotor treatments, and immunological products. However, insufficient information is available to identify individual compounds and usage within each of these groups.

2.3   PATHWAYS TO THE ENVIRONMENT

Veterinary medicines enter the environment by a number of different pathways. Currently the environmental risk assessment of veterinary medicinal products is only concerned with emission at or after use of the product (i.e., application and excretion; Montforts 1999). However, emissions may occur at any stage in a product's life cycle, including during production and during the disposal of the unused drugs, containers, and waste material containing the product (e.g., manure, fish water, and other dirty water; Montforts 1999). A summary of the possible emission routes to the environment is given below. The importance of individual routes into the environment for different types of medicine will vary according to the type of treatment, the route of administration, and the type of animal being treated.

2.3.1   EMISSIONS DURING MANUFACTURING AND FORMULATION

During the manufacture of an active pharmaceutical ingredient (API) and formulation of the finished drug product, raw materials, intermediates, and/or the active substance may be released to the air, to water in wastewater, and to land in the form of solid waste. During manufacture, the main route of release of drugs into the environment is probably via process waste effluents produced during the cleaning of API and manufacturing equipment used for coating, blending, tablet compressing, and packing (Velagaleti et al. 2002). Biological and chemical degradation processes such as biotransformation, mineralization, hydrolysis, and photolysis are thought to remove most drug residues before process waste effluents or sludge solids are discharged to surface waters or sewage treatment works or are released to land (Velagaleti et al. 2002). In addition, a number of practices are often implemented by industry to reduce waste generation and material losses.

These include process optimization, production scheduling, materials tracking, and waste stream segregation (US Environmental Protection Agency [USEPA] 1997). Losses to the environment arising during the manufacture or formulation of veterinary medicinal products are likely to be minimal.

Manufacturing plants employ several treatment methodologies and technologies to control and treat emissions and minimize the amount of waste produced. These include the use of condensers, scrubbers, adsorbent filters, and combustion or incineration for recovery and removal in air emissions. Neutralization, equalization, activated sludge, primary clarification, multimedia filtration, activated carbon, chemical oxidation, and advanced biological processes may be used for treatment of wastewaters (USEPA 1997).

2.3.2 Aquaculture

Chemotherapeutic medicines used in fish farming are limited to anti-infective agents for parasitic and microbial diseases, anesthetic agents, and medical disinfectants. Drugs are commonly administered as medicated feed, by injection, or, in the case of topical applications, as a bath formulation. Bacterial infections in fish are usually treated using medicated food pellets that are added directly to pens or cages (Samuelsen et al. 1992; Hektoen et al. 1995).

When infected, cultured fish show reduced appetite and thus feed intake. Consequently, a large proportion of medicated feed is not eaten, and this passes through the cages and is available for distribution to other environmental compartments. Furthermore, the bioavailability of many antibacterial agents is relatively low, and drugs may also enter the environment via feces and urine (Björklund and Bylund 1991; Hustvedt et al. 1991). In recent years, improved husbandry practices have reduced the amount of waste feed generated, and more recently authorized medicines have greater bioavailability ($F > 95\%$). Nevertheless, deposition of drugs from uneaten feed or feces on, or in, under-cage sediment can be a major route for environmental contamination by medicines used in aquaculture (Jacobsen and Berglind 1988; Björklund et al. 1991; Lunestad 1992). Once present on or in sediment, compounds may also leach back into the water column. During periods of treatment, some of the drugs entering the environment in waste feed and feces are also taken up by exploitative wild fish, shellfish, and crustaceans (Björklund et al. 1990; Samuelsen et al. 1992; Ervik et al. 1994; Capone et al. 1996).

When topical applications of chemotherapeutants are made, fish are usually crowded into a small water volume for treatment (Grave et al. 1991; Burka et al. 1997). Concentrated drugs are added directly to the water of open net pens or ponds, net pens enclosed by a tarpaulin, or tanks. Waste effluent is then either released into the surrounding water column or subject to local wastewater treatment and recycling (filters, settlement basins, and ponds; Grave et al. 1991; Burka et al. 1997; Montforts 1999). In addition, sludge recovered from wastewater-recycling activities may be applied directly to land or sold as fertilizer (Montforts 1999).

2.3.3   AGRICULTURE (LIVESTOCK PRODUCTION)

Large quantities of animal health products are used in agriculture to improve animal care and increase production. Some drugs used in livestock production are poorly absorbed by the gut, and the parent compound or metabolites are known to be excreted in the feces or urine, irrespective of the method of application (Campbell et al. 1983; Donoho 1987; Magnussen et al. 1991; Stout et al. 1991; Sommer et al. 1992). During livestock production, veterinary drugs enter the environment through removal and subsequent disposal of waste material (including manure or slurry and "dirty" waters), via excretion of feces and urine by grazing animals, through spillage during external application, via washoff from farmyard hard surfaces (e.g., concrete), or by direct discharge to the environment.

With all hormones, antibiotics, and other pharmaceutical agents administered either orally or by injection to animals, the major route of entry of the product into the environment is probably via excretion following use and the subsequent disposal of contaminated manure onto land (Halling-Sørensen et al. 2001). Many intensively reared farm animals are housed indoors for long periods at a time. Consequently, large quantities of farmyard manure, slurry, or litter are produced, which are then disposed of at relatively high application rates onto land (ADAS 1997, 1998; Montforts 1999). Although each class of livestock production has different housing and manure production characteristics, the emission and distribution routes for veterinary medicines are essentially similar. As well as contaminating the soil column, it is possible for veterinary medicines to leach to shallow groundwater from manured fields or even reach surface water bodies through surface runoff (Nessel et al. 1989; Hirsch et al. 1999; Hamscher et al. 2000a, 2000b, 2000c; Meyer et al. 2000). In addition, drugs administered to grazing animals or animals reared intensively outdoors are deposited directly to land or to surface water in dung or urine, exposing soil organisms to high local concentrations (Sommer and Overgaard Nielsen 1992; Strong 1992, 1993; McCracken 1993; Sommer et al. 1993; Strong and Wall 1994; Halling-Sørensen et al. 1998; Montforts 1999).

Another significant route for environmental contamination is the release of substances used in topical applications. Various substances are used externally on animals and poultry for the treatment of external or internal parasites and infection. Sheep in particular suffer from a number of external insect parasites for which treatment and protection are sometimes obligatory. The main methods of external treatment include plunge dipping or sheep dipping; pour-on formulations; and the use of showers or jetters. With all externally applied veterinary medicines, both diffuse and point source pollution can occur. Sheep-dipping activities provide several routes for environmental contamination. In dipping practice, chemicals may enter watercourses through inappropriate disposal of used dip, through leakage of used dip from dipping installations, and from excess dip draining from treated animals. Current disposal practices rely heavily on spreading used dip onto land (Health and Safety Executive 1997; Ministry of Agriculture Fisheries and Food 1998).

Washoff of chemicals from the fleeces of recently treated animals to soil, water, and hard surfaces may occur on the farm, during transport, or at stock markets. Some market authorities insist animals are dipped before entering the market to restrict the spread of disease, thus creating the potential for contaminated runoff from uncovered standing areas (Armstrong and Philips 1998). Medicines washed off, excreted, or spilled onto farmyard hard surfaces may be washed off to surface waters during periods of rainfall.

Other major sources of pollution arising from sheep dip chemicals are emissions from wool-washing plants and fellmongers (the initial processing stage of leather production; Armstrong and Philips 1998). Monitoring data (Environment Agency 1998) have demonstrated high numbers of Environmental Quality Standard (EQS) failures in the Yorkshire, United Kingdom area associated with the textile industry. Although effluent produced from the wool-washing process is normally treated for the removal of pollutants, this process is not always adequately effective, and chemicals may be released in discharges from the treatment plants. In addition, spills and leaks of untreated effluent directly to surface water drains from both fellmongers and wool treatment plants can occur (Environment Agency 1999).

The Environment Agency, working in partnership with representatives from the Scottish Environment Protection Agency, VMD, National Office of Animal Health, water companies, the textile industry, and sheep farmers, has produced a strategy for reducing sheep dip chemical pollution from the textile industry that provides detailed discussion and makes recommendations for dealing with the problem (Environment Agency 1999).

Other topically applied veterinary medicines likely to wash off following use include udder disinfectants from dairy units and endectocides for treating cattle parasites. Udder washings containing anti-infective agents and contaminated dirty water produced by dairy units may enter the environment through soakaways and surface water drains or via its inclusion in stored slurry and subsequent application to land. Washoff from the coats and skin of cattle treated with pour-on formulations can occur where the animals are exposed to rain shortly after dosing (Bloom and Matheson 1993). Residues of drugs in washoff may accumulate in localized high concentrations on land with high stocking densities. Contaminated surface runoff from open cattle yards (dirty water) is often collected and subsequently spread onto land. In addition, residues may wash off the backs and coats of grazing animals such as cattle and sheep that have access to surface water bodies as drinking water.

2.3.4 COMPANION AND DOMESTIC ANIMALS

To date, the environmental fate of veterinary medicines used in companion animals (pets) has not been extensively researched. This is probably because unlike production animals reared in agriculture, companion animals are kept on a small-scale basis and are therefore not subject to mass medication. Where used, drugs are likely to be dispersed into the environment via runoff or leaching from on-ground fecal material (Daughton and Ternes 1999). In addition, ectoparasiticides applied externally to canine species may contaminate surface water through direct loss from the coat if the animal enters the water.

2.3.5 Disposal of Unwanted Drugs

Veterinary medicines may be subject to disposal at any stage during their life cycle. It is probably fair to assume that, as with human medicines, a proportion of all prescribed or nonprescribed veterinary medicines will be unused and unwanted by the end user. The principal end users of veterinary medicines are veterinarians, livestock producers, and domestic users. Disposal of veterinary medicines by end users should be interpreted to include damaged, outdated, or outmoded animal medicines, as well as used containers and packages, contaminated sharps, applicators, and protective clothing (Cook 1995). Users are advised always to follow advice on the label regarding disposal and never to dispose of such items with domestic rubbish or down the drain or toilet.

Where appropriate, product label and safety data sheets provided by manufacturers provide information relating to the safe disposal of veterinary medicines and packaging. Distributors, veterinary practices, farmers, and feed compounders can also contact the manufacturer or local authority for advice, especially where large quantities of animal medicines require disposal and collection services are operated by some local authorities for the periodic disposal of special waste (Cook 1995). Users of companion animal products may return unwanted or unused product to the veterinarian or local pharmacist.

In practice, methods for disposal include flushing down the toilet, incineration, and local domestic waste collection. Domestic users will undoubtedly flush unwanted medicines down toilets or place them with the domestic refuse (Daughton and Ternes 1999). For ectoparasiticides, and in particular for sheep dips, containers should be returned to suppliers for correct disposal via high-temperature incineration or licensed landfill. In the United Kingdom, if on-farm disposal is planned, containers (water-soluble preparations) should be triple-rinsed before burning or burial away from watercourses or any land drains, as specified by the 1998 Code of Good Agricultural Practice for the Protection of Water. Inappropriate disposal of empty containers and unwanted product by careless operators may lead to contamination of soil and waters.

Unwanted or expired products that are returned to the manufacturer are usually disposed of through incineration or landfilling at suitable sites (Velagaleti et al. 2002). Where medicines are disposed of in sufficient quantities to unlined landfill sites, residues present in the leachate may reach shallow groundwater and surface waters (Holm et al. 1995).

2.4 SUMMARY

The impact of veterinary medicines on the environment will depend on several factors, including the amounts used, animal husbandry practices, treatment type and dose, metabolism within the animal, method of administration, environmental toxicity, physicochemical properties, soil type, weather, manure storage and handling practices, and degradation rates in manure and slurry.

The importance of individual routes into the environment for different types of veterinary medicines will vary according to the type of treatment and livestock category. Treatments used in aquaculture have a high potential to reach the aquatic environment. The main routes of entry to the terrestrial environment will be from the use of veterinary medicines in intensively reared livestock, via the application of slurry and manure to land, and via the use of veterinary medicines in pasture-reared animals where residues from medicines will be excreted directly into the environment. Veterinary medicines applied to land by spreading of slurry may also enter the aquatic environment indirectly via surface runoff or leaching to groundwater. It is likely that topical treatments will have a greater potential to be released to the environment than treatments administered orally or by injection. Inputs from the manufacturing process, companion animal treatments, and disposal are likely to be minimal in comparison.

This chapter has reviewed the data available in the public domain on veterinary medicine's usage and pathways to the environment. Although there is a large body of data available, there are clearly several gaps in these data and in our understanding of the impacts of veterinary medicines on the environment. On the basis of this review, the following gaps can be identified.

1) Usage data are unavailable for many groups of veterinary medicines and for several geographical regions, which makes it difficult to establish whether these substances pose a risk to the environment. It is, therefore, recommended that usage information be obtained for these groups, including the antiseptics, steroids, diuretics, cardiovascular and respiratory treatments, locomotor treatments, and immunological products. Better usage data will assist in designing more robust hazard and risk management strategies that are tailored to geographically explicit usage patterns.

2) From the information available, it appears that inputs from aquaculture and herd or flock treatments are probably the most significant in terms of environmental exposure. This is mainly because many aquaculture treatments are dosed directly into the aquatic environment, and herd or flock treatments may be excreted directly onto pasture. However, the relative significance of novel routes of entry to the environment from livestock treatments, such as washoff following topical treatment and farmyard runoff, and aerial emissions, has not generally been considered. For example, the significance of exposure to the environment from the disposal of used containers or from discharge from manufacturing sites should be investigated further. In addition, substances may be

released to the environment as a result of off-label use and poor slurry management practice. The significance of these exposure routes is currently unknown.

REFERENCES

ADAS. 1997. Animal manure practices in the pig industry: survey report. July. Wolverhampton (UK): ADAS Consulting Ltd.

ADAS. 1998. Animal manure practices in the dairy industry: survey report. March. Wolverhampton (UK): ADAS Consulting Ltd.

Armstrong A, Philips K. 1998. A strategic review of sheep dipping. R&D Technical Report P170. Bristol (UK): Environmental Agency.

Björklund HV, Bondestam J, Bylund G. 1990. Residues of oxytetracycline in wild fish and sediments from fish farms. Aquaculture 86:359–367.

Björklund HV, Bylund G. 1991. Comparative pharmacokinetics and bioavailability of oxolinic acid and oxytetracycline in rainbow trout (*Oncorhynchus mykiss*). Xenobiotica 21:1511–1520.

Björklund HV, Råbergh CMI, Bylund G. 1991. Residues of oxolinic acid and oxytetracycline in fish and sediments from fish farms. Aquaculture 97:85–96.

Bloom RA, Matheson JC. 1993. Environmental assessment of avermectins by the US Food and Drug Administration. Vet Parasitol 48:281–294.

Boxall ABA, Fogg LA, Kay P, Blackwell PA, Pemberton EJ, Croxford A. 2004. Veterinary medicines in the environment. Revs Environ Contam Toxicol 180:1–91.

Boxall ABA, Sherratt T, Pudner V, Pope L. 2007. A screening level model for assessing the impacts of veterinary medicines on dung organisms. Environ Sci Tech 41(7):2630–2635.

Burka JF, Hammell KL, Horsberg TE, Johnson GR, Rainnie DJ, Speare DJ. 1997. Drugs used in salmonid aquaculture: a review. J Vet Pharmacol Therapeut 20:333–349.

Campbell WC, Fisher MH, Stapley EO, Albers-Schonberg G, Jacob TA. 1983. Ivermectin: a potent new antiparasitic agent. Science 221:823–828.

Capone DG, Weston DP, Miller V, Shoemaker C. 1996. Antibacterial residues in marine sediments and invertebrates following chemotherapy in aquaculture. Aquaculture 145:55–75.

Cook RR. 1995. Disposal of animal medicines. In: Animal medicines a user's guide: transport, storage and disposal of animal medicines. Enfield (UK): National Office of Animal Health Ltd.

Daughton CG, Ternes TA. 1999. Pharmaceuticals and personal care products in the environment: agents of subtle change? Special report. Environ Health Perspect Suppl 107:907–938.

Davies IM, Gillibrand PA, McHenry JG, Rae GH. 1998. Environmental risk of ivermectin to sediment-dwelling organisms. Aquaculture 163:29–46.

Donoho AL. 1987. Metabolism and residue studies with actaplanin. Drug Metabol Rev 18:163–176.

Environment Agency. 1997. The occurrence of sheep-dip pesticides in environmental waters. Peterborough (UK): National Centre for Toxic and Persistent Substances, Environment Agency.

Environment Agency. 1998. Pesticides 1998: a summary of monitoring of the aquatic environment in England and Wales. Wallingford (UK): National Centre for Ecotoxicology and Hazardous Substances, Environment Agency.

Environment Agency. 1999. Sheep dip chemicals and textiles working group: a strategy for reducing sheep dip chemical pollution from the textile industry. Wallingford (UK): National Centre for Ecotoxicology and Hazardous Substances, Environment Agency.

Environment Agency. 2000. Welsh sheep dip monitoring programme 1999. Cardiff (UK): Environment Agency Wales.

Ervik A, Thorsen B, Eriksen V, Lunestad BT, Samuelsen OB. 1994. Impact of administering antibacterial agents on wild fish and blue mussels *Mytilus edulis* in the vicinity of fish farms. Diseases Aquat Organisms 18:45–51.

Grave K, Engelstad M, Søli NE, Toverud EL. 1991. Clinical use of dichlorvos (Nuvan®) and trichlorfon (Neguvon®) in the treatment of salmon louse, *Leptophtheirus salmonis*: compliance with the recommended treatment procedures. Acta Vet Scand 32:9–14.

Gustafson RH, Bowen RE. 1997. Antibiotic use in animal agriculture. J Appl Microbiol 83:531–541.

Halling-Sørensen B, Jensen J, Tjørnelund J, Montforts MHMM. 2001. Worst-case estimations of predicted environmental soil concentrations (PEC) of selected veterinary antibiotics and residues used in Danish agriculture. In: Kümmerer K, editor. Pharmaceuticals in the environment: sources, fate, effects and risks. Berlin (Germany): Springer.

Halling-Sørensen B, Nors Nielsen S, Lanzky PF, Ingerslev F, Holten Lützhøft HC, Jørgensen SE. 1998. Occurrence, fate and effect of pharmaceutical substances in the environment: a review. Chemosphere 36:357–393.

Hamscher G, Abu-Quare A, Sczesny S, Höper H, Nau H. 2000a. Determination of tetracyclines and tylosin in soil and water samples from agricultural areas in lower Saxony. In: van Ginkel LA, Ruiter A, editors. Proceedings of the Euroresidue IV conference, Veldhoven, Netherlands, 8–10 May. Bilthoven (The Netherlands): National Institute of Public Health and the Environment (RIVM).

Hamscher G, Sczesny S, Abu-Quare A, Höper H, Nau H. 2000b. Substances with pharmacological effects including hormonally active substances in the environment: identification of tetracyclines in soil fertilised with animal slurry. Dtsch tierärztl Wschr 107:293–348.

Hamscher G, Sczesny S, Höper H, Nau H. 2000c. Tetracycline and chlortetracycline residues in soil fertilized with liquid manure. Proceedings of Livestock Farming and the Environment, Hannover, Germany, 28–29 September.

Health and Safety Executive. 1997. Sheep dipping. Booklet AS29 (Rev. 2). Bootle (UK): Health and Safety Executive.

Hektoen H, Berge JA, Hormazabal V, Yndestad M. 1995. Persistence of antibacterial agents in marine sediments. Aquaculture 133:175–184.

Hirsch R, Ternes T, Haberer K, Kratz KL. 1999. Occurrence of antibiotics in the aquatic environment. Sci Total Environ 225:109–118.

Holm JV, Rügge K, Bjerg PL, Christensen TH. 1995. Occurrence and distribution of pharmaceutical organic compounds in the groundwater down gradient of a landfill (Grinsted, Denmark). Environ Sci Technol 29:1415–1420.

Hustvedt SO, Salte R, Kvendseth O, Vassvik V. 1991. Bioavailability of oxolinic acid in Atlantic salmon (*Salmo salar* L) from medicated feed. Aquaculture 97:305–310.

Jacobsen P, Berglind L. 1988. Persistence of oxytetracycline in sediments from fish farms. Aquaculture 70:365–370.

Jørgensen SE, Halling-Sørensen B. 2000. Drugs in the environment. Chemosphere 40:691–699.

Kools SAE, Moltmann JF, Knacker T. 2008. Estimating usage of veterinary medicines in the European Union. Regulatory Toxicology and Pharmacology.

Liddel JS. 2000. Sheep ectoparasiticide use in the UK: 1993, 1997 and 1999. Paper presented to the 5th International Sheep Veterinary Congress, Stellenbosch, South Africa.

Lunestad BT. 1992. Fate and effects of antibacterial agents in aquatic environments. Proceedings of the Conference on Chemotherapy in Aquaculture: from theory to reality. Paris (France): Office International des Epizooties.

Magnussen JD, Dalidowicz JE, Thomson TD, Donoho AL. 1991. Tissue residues and metabolism of avilamycin in swine and rats. J Agric Food Chem 39:306–310.

McCracken DI. 1993. The potential for avermectins to affect wildlife. Vet Parasitol 48:273–280.

McKellar QA. 1997. Ecotoxicology and residues of anthelmintic compounds. Vet Parasitol 72:413–435.

Meyer MT, Bumgarner JE, Varns JL, Daughtridge JV, Thurman EM, Hostetler KA. 2000. Use of radioimmunoassay as a screen for antibiotics in confined animal feeding operations and confirmation by liquid chromatography/mass spectrometry. Sci Total Environ 248:181–187.

Ministry of Agriculture Fisheries and Food. 1998. Code of good agricultural practice for the protection of water. Cardiff (UK): Ministry of Agriculture Fisheries and Food, Welsh Office Agricultural Department.

Montforts MHMM. 1999. Environmental risk assessment for veterinary medicinal products part 1: other than GMO-containing and immunological products. RIVM report 601300 001, April. Bilthoven (The Netherlands): National Institute of Public Health and the Environment.

Nessel RJ, Wallace DH, Wehner TA, Tait WE, Gomez L. 1989. Environmental fate of ivermectin in a cattle feedlot. Chemosphere 18:1531–1541.

Pepper T, Carter A. 2000. Monitoring of pesticides in the environment. Report prepared for the Pesticides in the Environment Working Group, R&D Project E1-076. Bristol (UK): Environment Agency.

Ridsdill-Smith TJ. 1988. Survival and reproduction of *Musca vetustissima* Walker (*Diptera muscidae*) and a scarabaeine dung beetle in dung cattle treated with avermectin B1. J Aust Entomol Soc 27:175–178.

Samuelsen OB, Lunestad BT, Husevåg B, Hølleland T, Ervik A. 1992. Residues of oxolinic acid in wild fauna following medication in fish farms. Diseases Aquat Organisms 12:111–119.

Sarmah AK, Meyer MT, Boxall ABA. 2006. A global perspective on the use, sales, exposure pathways, occurrence, fate and effects of veterinary antibiotics (VAs) in the environment. Chemosphere 65:725–759.

Scottish Environment Protection Agency. 2000. Long-term biological monitoring trends in the Tay System 1988–1999. Stirling (UK): Scottish Environment Protection Agency, Eastern Region.

Sommer C, Gronvold J, Holter P, Nansen P. 1993. Effect of ivermectin on two afrotropical beetles *Onthophagus gazella* and *Diastellopalpus quinquedens* (Coleoptera: Scarabaeidae). Vet Parasitol 48:171–179.

Sommer C, Overgaard Nielsen B. 1992. Larvae of the dung beetle *Onthophagus gazella* F. (Col., Scarabaeidae) exposed to lethal and sublethal ivermectin concentrations. J Appl Entomol 114:502–509.

Sommer C, Steffansen B, Overgaard Nielsen B, Grønvold J, Kagn-Jensen KM, Brøchner Jespersen J, Springborg J, Nansen P. 1992. Ivermectin excreted in cattle dung after sub-cutaneous injection or pour-on treatment: concentrations and impact on dung fauna. Bull Entomol Res 82:257–264.

Stout SJ, Wu J, da Cunha AR, King KG, Lee A. 1991. Maduramycin α: characterisation of 14C-derived residues in turkey excreta. J Agric Food Chem 39:386–391.

Strong L. 1992. Avermectins: a review of their impact on insects of cattle dung. Bull Entomol Res 82:265–274.

Strong L. 1993. Overview: the impact of avermectins on pastureland ecology. Vet Parasitol 48:3–17.

Strong L, Wall R. 1994. Effects of ivermectin and moxidectin on the insects of cattle dung. Bull Entomol Res 84:403–409.

[USEPA] US Environmental Protection Agency. 1997. Profile of the pharmaceutical manufacturing industry. EPA/310-R-97-005. Washington (DC): USEPA Office of Compliance.

Velagaleti R., Burns PK, Gill M, Prothro J. 2002. Impact of current good manufacturing practices and emission regulations and guidances on the discharge of pharmaceutical chemicals into the environment from manufacturing, use and disposal. Environment Health Perspectives 110(3):213–220.

Veterinary Medicines Directorate. 2001. Sales of antimicrobial products used as veterinary medicines and growth promoters in the UK in 1999. Addlestone (UK): Veterinary Medicines Directorate.

Wall R, Strong L. 1987. Environmental consequences of treating cattle with the antiparasitic drug ivermectin. Nature 327:418–421.

3 Environmental Risk Assessment and Management of Veterinary Medicines

Joop de Knecht, Tatiana Boucard,
Bryan W. Brooks, Mark Crane, Charles Eirkson,
Sarah Gerould, Jan Koschorreck, Gregor Scheef,
Keith R. Solomon, and Zhixing Yan

3.1 INTRODUCTION

Although often considered as a single group of chemicals, veterinary medicines are a diverse group of different products containing a broad range of compounds belonging to different chemical classes and used for a diverse assortment of conditions (see Table 2.1 in Chapter 2). Antiparasiticides control external parasites such as ticks or sea lice (ectoparasiticides), internal parasites such as gastrointestinal worms and protozoans (endoparasiticides), or both (endectocides). Antibiotics are used for the treatment and prevention of bacterial infections, whereas fungicides are administered to treat fungal or yeast infestation. Hormones regulate growth, reproduction, and other bodily functions.

Veterinary medicines are used to treat many groups of animals, such as terrestrial and aquatic animals that are used for food, and companion animals. Taxonomically, the groups include mammals (e.g., cattle, horses, pigs, sheep, goats, dogs, and cats), birds (e.g., chickens and turkeys), fish, and invertebrates (e.g., bees, lobsters, and shrimps). This diverse group of animals necessitates a variety of treatment techniques. Veterinary medicines are administered orally, parenterally (intramuscular, intravenous, and subcutaneous injection), and topically (dip, spray, pour-on, spot-on, ear tag, collar, and aquaculture water baths). Veterinary medicines are not usually directly applied to the environment except for some aquaculture treatments, although manure, drainage from sheep dip, releases from aquaculture facilities, scavenging of carcasses, and other environmental releases result in environmental exposure to nontarget organisms.

Releases of veterinary medicines into the environment can take place at any step in the life cycle of the product. However, veterinary medicines have a carefully regulated, definable set of uses, resulting in restricted ranges of scenarios for environmental exposure. The dosage, route of application, type of target animals, excretion, metabolic and degradation products, route of entry into the environment, and agricultural practice determine the range of exposures. Premarket environmental risk assessment focuses on exposure during or after use of the product and considers a number of different exposure scenarios; appropriate mitigation measures follow from these factors (see Section 3.3). These exposure scenarios are as follows:

- Runoff during or following during external application
- Releases of veterinary medicine in waste material (manure, dirty drinking water, and aquaculture water) during cleanup, storage, removal, and land application
- Excretion via feces and urine (grazing animals)
- Spillage at external application site or direct exposure outdoors
- Disposal of containers (bottles and flea collars)

In contrast to most other chemicals, many veterinary medicines are defined by a specific biological activity intended to exert adverse effects on either eukaryotes (e.g., fungi, helminthes, and arthropods) or prokaryotes (e.g., bacteria). Their intended toxicity also results in a potential to cause toxic effects to nontarget species in the environment. Knowledge of the active substance's mode of action, derived from pharmacodynamic studies, could help to identify specific taxonomic groups for which an increased risk should be assessed. Also, information commonly used for the human health risk assessment, such as absorption, distribution, metabolism, and excretion of the compound (ADME), as well as its toxicity toward mammals, birds, and aquatic organisms (depending on the envisaged target and nontarget species) are useful information in the environmental risk assessment of veterinary medicines.

Compared to other chemicals, such as nonprescription drugs and high production volume (HPV) chemicals, veterinary medicines are used only in limited amounts. For example, the total usage of therapeutic antibiotics in the United Kingdom in 2004 amounted to 476 tons active ingredients (Veterinary Medicines Directorate [VMD] 2005), whereas in the year 2000, 12.7 tons of anthelmintics (active ingredient) were administered. In comparison, the total amount of pesticides used in the United Kingdom in 2004 amounted to 26 356 tons of active ingredient (European Crop Protection Association n.d.; see http://www.ecpa.be), whereas 7188 tons of the HPV chemical nonylphenol were estimated to be used in the United Kingdom in 1997 (Defra 2004). Even if the overall usage of veterinary medicines is relatively small compared to that of other chemicals, the potential for adverse nontarget effects makes a thorough environmental risk assessment necessary.

Like other medicinal products, the packaging insert and text on the label provide clear instructions for the use of the veterinary medicine (see Section 3.3).

Depending on the outcome of the regulatory environmental risk assessment associated with marketing authorization, in addition to the standard information, the label might contain specific remarks related to risk measurements and/or mitigation as well as warning statements related to environmental safety and disposal. Products may require a prescription or administration by a professional user, such as a veterinarian or farmer. Veterinary medicines that cause greatest concerns with respect to safety, such as parasiticides and antibiotics in food animals, are often regulated in Europe by requiring their prescription by a veterinarian. These requirements may help to limit the risk of environmental exposure to the level identified as acceptable in the course of the risk assessment.

In the European Union, an initial marketing authorization for a veterinary medicine is valid for a period of 5 years only. After this period, the risk–benefit balance has to be reevaluated in a "renewal" by taking into account all new information received after placing the medicine on the market, in addition to any new regulatory requirements that have emerged. Once renewed, the marketing authorization is valid for an unlimited period; however, the regulatory bodies may require the applicant to submit documentation related to a medicine's quality, safety (including environmental safety), and efficacy at any time. A renewal procedure is not established in the United States, so regulatory bodies there can typically only require new environmental safety information related to the product when a supplemental authorization is being requested for changes in existing product conditions, such as a new marketing claim or disease indication.

3.2   VETERINARY MEDICINES IN REGULATORY PERSPECTIVE

3.2.1   LEGISLATION, SCOPE, AND PAST GUIDELINES FOR ENVIRONMENTAL RISK ASSESSMENT (ERA) OF VETERINARY MEDICINES

Over the last 2 decades, the environmental safety of medicinal products has gained increasing prominence not only in the scientific community but also in the public's perception. Pharmaceutical companies and regulatory bodies have reacted to this by assessing the potential environmental risk arising from the use and the disposal of medicines prior to marketing. In the 1970s, the US Food and Drug Administration (FDA) began requiring an environmental risk assessment for many new human and veterinary medicines. Other regions followed in the 1980s (Australia for veterinary medicines) and 1990s (the European Union and Canada for both veterinary and human medicines). Japan has prepared a regulatory framework for veterinary medicines. From an environmental perspective and on a worldwide scale, more attention is currently given to the safety of veterinary medicines than to the potential environmental risks of human medicines: both the legal requirements and the concepts guiding the risk assessment are more stringent for assessing environmental risks.

Table 3.1 summarizes the current regulatory situation for assessing the environmental risks of veterinary medicines in several important jurisdictions, as discussed further below.

TABLE 3.1

Overview of the regulatory situation for environmental risk assessment of veterinary medicines

Region	Regulatory agency	Legal requirements	ERA guidelines
European Union	Member State specific, European Medicines Agency	Directive 2004/28/EC (European Parliament 2004b) Regulation EC/726/2004 (European Parliament 2004c)	VICH phase I (2000) VICH phase II (2005)
United States	Food and Drug Administration Center for Veterinary Medicine	Federal Food, Drug and Cosmetic Act National Environmental Policy Act	VICH phase I (1998) VICH phase II (2006)
Japan	Ministry of Agriculture, Forestry and Fisheries	Expected in 2006	VICH phase I and II ongoing in 2008
Australia	Pesticides and Veterinary Medicines Authority Department of Environment	Agricultural and Veterinary Chemicals Code Act (1994)	VICH phase I (July 2001), Veterinary Manual of Data Requirements and Guidelines (1997)
Canada	Environmental Assessment Unit of Health Canada	New Substances Notification Regulations of the Canadian Environmental Protection Act	So far, environmental risk assessment related to assessment of chemicals

3.2.1.1   United States

The FDA is responsible for the market authorization of medicines. The requirement to submit environmental impact information (Code of Federal Regulations title 21, part 25 [21 CFR25]; see National Archives and Records Administration 2004) was issued in 1973. In practice, the FDA began asking companies to submit reports on environmental risk in the late 1980s. The National Environmental Policy Act of 1969 (NEPA) requires an assessment of the potential environmental impact of a medicine's proposed use but does not necessarily require the FDA to take the most environmentally beneficial action. An environmental review by the FDA can comprise 1) granting a categorical exclusion for approval actions on veterinary medicines that are not expected to significantly impact the environment, 2) an environmental assessment (EA) for approval actions that are not categorically excluded to determine whether a veterinary medicine may significantly impact the environment, or 3) an environmental impact statement (EIS) for approval actions on veterinary medicines that may significantly impact the environment.

For veterinary medicines, there are a number of approval actions that are generally eligible for a categorical exclusion unless extraordinary circumstances exist. These include the following:

- Applications for new drugs to be used in nonfood animals
- Applications for new drugs for minor species, including wildlife and endangered species, when previously approved under similar animal management practices
- Applications for new therapeutics to be used under veterinarian order or prescription in terrestrial species, unless the 100 ppb criteria specified in the VICH phase I guidance (International Cooperation on Harmonization of Technical Requirements for Registration of Veterinary Products [VICH] 2002) is exceeded
- New drug applications for substances that occur naturally in the environment when the use will not alter the concentration or distribution of the substance (or its metabolites or degradation products) in the environment
- New and supplemental animal drug applications when the approval will not increase the use of the drug (e.g., minor formulation changes, combinations of previously approved drugs, and generic copies of pioneer drugs)

For the environmental impact assessment of veterinary medicinal products, VICH phase I and phase II assessments (see Section 3.2.2) have been implemented in the US regulatory scheme. These assessments are incorporated into an environmental assessment document that determines whether an environmental impact statement needs to be prepared. If not, a finding of no significant impact (FONSI) is issued by the FDA. Sometimes the FONSI may include risk management or mitigation measures that are used to avoid or reduce environmental impacts.

3.2.1.2   European Union

In Europe there are 2 types of authorizations. In a centralized procedure a product is authorized by the European Medicines Agency in all EU member states simultaneously. In contrast, a national authorization is acquired from the regulatory body of an individual member state by a strictly national procedure, a mutual recognition, or a decentralized procedure. The authorization process is strictly harmonized between the 27 EU member states by EU Directives and Regulations. The need to demonstrate the environmental safety of veterinary and human medicines was established in 1990 (by EU Directive 90/676/EEC) and 1993 (EU Directive 93/39/EEC), respectively. Directives 2004/27/EC (on human medicines; European Parliament 2001b, 2004a) and 2004/28/EC (on veterinary medicines; European Parliament 2001a, 2004b) introduced a definition for the risk of a medicinal product relating to its quality, safety, efficacy, and undesirable environmental effects.

For veterinary medicines the risk–benefit analysis, which is the evaluation of positive therapeutic effects of a medicinal product in relation to risks, includes any environmental risks. In contrast, the overall benefit of human medicines is stressed by excluding environmental concerns from the risk–benefit analysis. The granting of a marketing authorization for a veterinary medicinal product may therefore be refused due to an unacceptable risk to the environment, although this cannot occur for human medicines. Both the human and the veterinary community codes aim at risk mitigation measures via labeling to reduce any environmental risks arising from the use of a product.

In 1996 the Committee for Veterinary Medicinal Products of the European Agency for the Evaluation of Medicinal Products adopted a Note for Guidance for the evaluation of the environmental risk assessment for veterinary medicinal products (European Agency for the Evaluation of Medicinal Products 1998). This document has now been replaced by VICH phase I and phase II in 2000 and 2004, respectively, which are discussed further below in Section 3.2.2.

3.2.1.3 Japan

So far the environmental risk assessment of veterinary and human medicines has not been established in Japanese regulations. A regulation is expected to be released by the Ministry of Agriculture, Forestry and Fisheries for environmental risk assessment of veterinary medicines, but it has not yet been decided whether the new regulation will include risk mitigation measures. Japan took part in the tripartite elaboration of the VICH phase I and II documents (VICH 2002, 2004), which came into force in 2007. Guidelines for the exposure estimation to go along with the VICH documents will be developed.

3.2.1.4 Australia

The authorization of veterinary medicines falls under the Australian Pesticides and Veterinary Medicines Authority. The Department of Environment began assessing the environmental risk for pesticides and veterinary medicines in 1986. The current legal basis is the Agricultural and Veterinary Chemicals Code Act (1994; Commonwealth of Australia 2005), which requires that the use of a proposed veterinary medicinal product would not be likely to have an unintended effect that is harmful to animals, plants, or the environment. Label restrictions and warning statements are mentioned in the legal text to mitigate an environmental risk, and a serious environmental risk can lead to the denial of the marketing authorization.

Guidance on environmental risk assessment was given in 1997 in the Veterinary Manual of Data Requirements and Guidelines. As in the European Union and the United States, VICH phase I came into force in July 2001 (with some qualifications). VICH phase II has become part of the Veterinary Manual of Data Requirements and Guidelines in the near future.

3.2.1.5 Canada

The Canadian Food and Drugs Act currently regulates all new substances in human and veterinary medicine products prior to import or sale. The Canadian Environmental Protection Act (1999) established the need for an environmental risk assessment under the New Substances Notification Regulations prior to manufacture or import. The environmental risk assessments for medicines are carried out by the Environmental Assessment Unit of Health Canada. Data requirements are triggered by estimated sales volumes.

No specific guidelines for the evaluation of the environmental risks of human or veterinary medicinal products have been established so far. However, Health Canada has initiated a consultative process to determine the most appropriate regulations for veterinary medicines. The Government of Canada will make every possible effort to incorporate the requirements defined in the VICH Ecotoxicity Guideline in the development of the Environmental Assessment Regulations. This approach is commensurate with the Canadian Veterinary Drugs Directorate's efforts toward international harmonization and to its participation in VICH.

3.2.2   Current Guidelines: VICH and the VICH–EU Technical Guidance Document (VICH–EU–TGD)

In order to achieve harmonization between Europe, the United States, Japan, Canada, and Australia and New Zealand on the data requirements for the registration of veterinary medicines, the VICH Steering Committee (VICH SC) authorized in 1996 the formation of a working group to develop a 2-phased, logically tiered approach outlined in 2 guidelines (phase I and phase II) for the environmental risk assessment of veterinary medicines. The working group had a single industry and a single regulatory representative from each of the regions. The VICH guidance documents on phase I and phase II were finalized in June 2000 and October 2004, respectively.

The VICH phase I makes use of a decision tree (Figure 3.1), which applicants work through until they are able to determine whether or not their product qualifies for a phase II assessment. In principle, exemption from further testing in both phases I and II is in principle acceptable for the following:

- Natural substances, the use of which will not alter the concentration or distribution of the substance in the environment, such as vitamins, electrolytes, proteins, and peptides.
- Products intended for administration to nonfood animals (with varying definition of nonfood animals in the VICH regions).
- Veterinary medicines that are already approved for use in a major species, provided that the minor species is reared and treated similarly to the major species.
- Products used to treat a small number of animals in a flock or herd.
- Veterinary medicines that are extensively metabolized in the treated animal. A medicine may be defined as "extensively metabolized" when analysis of excreta shows that it is converted into metabolites that have lost structural resemblance with the parent compound or are common to basic biochemical pathways, or when no single metabolite or the parent medicine exceeds 5% of the total radioactivity excreted.

Phase I is then further divided by an assessment for veterinary medicines used into the so-called aquatic and terrestrial branches. In the aquatic branch, any veterinary medicine intended for use in open systems is directed to phase II if the concentration in effluent from an aquaculture facility is predicted to be

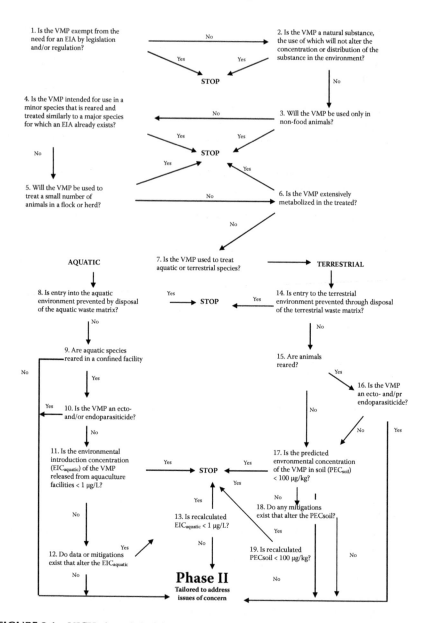

FIGURE 3.1   VICH phase 1 decision tree.

greater than 1 µg L⁻¹. In the terrestrial branch, veterinary medicines that are endo- and ectoparasiticides used in pasture will be advanced automatically to phase II because they are pharmacologically active against organisms that are biologically related to pasture invertebrates. For all other veterinary medicines, phase II assessment is required only if the predicted environmental concentration (PEC) in soil is greater than 100 µg kg⁻¹.

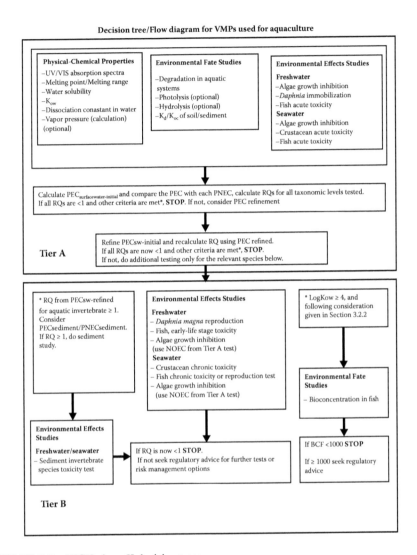

FIGURE 3.2   VICH phase II decision trees.

The VICH phase II guidance includes sections and decision trees for each of the major branches: 1) aquaculture, 2) intensively reared terrestrial animals, and 3) pasture animals (Figure 3.2). The trees include specific decision-making criteria appropriate to each branch. The guidance includes 2 tiers (tier A and tier B), for which there are OECD or International Standards Organization (ISO) data requirements for physical and chemical properties, environmental fate, and environmental effects testing (Table 3.2).

All testing is carried out on the active ingredient based on a total residue approach, and assuming that any metabolites are either equally or less toxic than the active ingredient. The possible exception to this is veterinary medicines such

TABLE 3.2

International Cooperation on Harmonization of Technical Requirements for Registration of Veterinary Products (VICH) tier A fate and effects studies to be included

Studies	Guideline
Fate and behavior	
Soil adsorption/desorption	OECD 106
Soil biodegradation (route and rate)	OECD 307
Degradation in aquatic systems	OECD 308
Photolysis (optional)	Seek regulatory guidance
Hydrolysis (optional)	OECD 111
Aquatic effects	
Algal growth inhibition	OECD 201 (FW) ISO 10253 (SW)
Daphnia immobilization	OECD 202 (FW) ISO 14669 (SW)
Fish acute toxicity	OECD 203
Terrestrial effects	
Nitrogen transformation (28 days)	OECD 216
Terrestrial plants	OECD 208
Earthworm subacute/reproduction	OECD 220/222
Dung fly larvae	No guideline available
Dung beetle larvae	No guideline available

Note: FW: freshwater; SW: saltwater.

[a] For substances with antimicrobial activity, some regulatory authorities prefer testing a blue-green alga rather than a green alga.

as inactive pro-drugs that are quickly and efficiently metabolized into an active drug, when it may be more appropriate to test the metabolite. Because the acute earthworm study was considered to be relatively insensitive, the VICH working group agreed instead to recommend a chronic earthworm study.

In principle, for all veterinary medicines used in intensively reared and pasture animals, all toxicity studies (both terrestrial and aquatic) are required, unless it can be argued that one of the compartments is not exposed. Toxicity studies for sediment-dwelling organisms are required when the PEC/PNEC for water column invertebrates is > 1.

The assessment in tier A starts with a PEC calculation based on the total residue. If the PEC/PNEC is ≥ 1, then available metabolism and excretion data from the residues part of the dossier should be considered to refine the PEC. Metabolites that represent 10% or more of the excreted dose and that do not form part of biochemical pathways should be summed to allow the PEC to be recalculated. In addition, the PEC may be refined further by several adjustments to account for processes such as the following:

- Degradation of the active ingredient and relevant metabolites during storage of manure before spreading on fields, as appropriate; and
- Degradation of the active ingredient and relevant metabolites in the field, using the results of the laboratory soil degradation study from tier A. Time to mineralization or degradation to substances that are part of biochemical pathways can be used to refine the PEC in this case.

The VICH phase II is based on a risk quotient (RQ) approach determined for every test species. If the RQ after PEC refinement is still > 1 for any of the species tested, then evaluation of the chemical moves to tier B and additional toxicity studies for the affected species are recommended (Table 3.3).

In tier A an assessment factor (AF) of 1000 is applied to endpoints from *Daphnia* and fish studies and an AF of 100 is applied to algal endpoints. An AF of 10 is used to derive a PNEC from chronic toxicity studies in tier B.

Risks to microorganisms are evaluated in the same manner as is currently done in risk assessment for the registration of pesticides. When the difference in rates of nitrate formation between the maximum PEC and control is < 25% at any sampling time after day 28, the medicine is considered to have no long-term influence on nitrogen transformation in soils. If this is not the case, the test should be extended to 100 days and evaluated in tier B.

For plants, an AF of 100 is applied to the lowest EC50 of 3 species tested. If the RQ > 1, the test should be repeated in tier B on 2 additional species from the most sensitive species category in the tier A test, in addition to repeating the test on the most sensitive species. The NOEC is then used to derive a PNEC by applying an AF of 10. Because in Tier A the effect on earthworms has already been tested in a reproduction study, the PNEC is derived from the NOEC by also applying an AF of 10.

TABLE 3.3
International Cooperation on Harmonization of Technical Requirements for Registration of Veterinary Products (VICH) tier B effects studies

Studies	Guideline
Bioconcentration in fish	OECD 305
Algae growth inhibition	OECD 201 (FW) and ISO 10253 (SW)
Daphnia magna reproduction	OECD 211
Fish, early life stage	OECD 210
Sediment invertebrate species toxicity	OECD 218 and 219
Nitrogen transformation (100 days; extension of tier A study)	OECD 216
Terrestrial plants growth, more species	OECD 208

Note: FW: freshwater; SW: saltwater.

For risk assessment in dung, the RQ is determined for dung fly larvae and dung beetle larvae, using an acute endpoint and an AF of 100.

Although not included in VICH phase I or II guidance, in the VICH–EU–TGD the following scenarios for secondary poisoning are also considered: 1) birds eating contaminated earthworms and 2) fish-eating predators eating fish that, in turn, eat small aquatic organisms that have accumulated the veterinary medicine. For birds exposed through sheep dips, the risk is assessed by using acute LD50 data, as chronic exposure through this route is unlikely.

Tests for toxicity to vertebrates (mammals and birds) are not recommended at tier A. However, the VICH working group recognized that there may be cases where there is both high toxicity and potential exposure through the food chain and therefore a consequent risk. An example of this is the risk to birds that feed on the backs of animals that have been treated with pour-on formulations of endo- and ectoparasiticides with potentially high mammalian and/or avian toxicity. In this case the applicant should consider the mammalian and (if available) avian toxicity data and seek regulatory guidance as to whether additional data are needed. Similarly if the log K_{ow} of a veterinary medicine is > 4, the risk of accumulation by earthworms and further biomagnification through the food chain should be considered.

Although not all taxonomic groups are tested, these measurement endpoints are thought to provide the necessary information to protect the functional and structural integrity of exposed ecosystems and to estimate adequately the risks to other aesthetically and commercially valuable organisms, such as butterflies, salmon, and eagles.

Several issues could not be harmonized during the VICH process. For example, default values and models for PEC calculation were considered to be regionally based and therefore outside the scope of VICH. These unresolved issues led to the conclusion by European regulators that there was a need for further guidance in Europe in the form of an EU–VICH–TGD. This contains guidance on the following issues:

- Default values for exposure calculation
- Exposure models for soil, leaching to groundwater, and runoff to surface waters
- Bioaccumulation and secondary poisoning
- Test strategies for dung fauna
- Groundwater assessment
- Higher tier studies for earthworms and plants
- Degradation of veterinary medicines during manure storage
- Data presentation and the structure of an expert report
- Risk mitigation measures
- Explanations and examples of the VICH approach

However, there remains no guidance on pharmacovigilance or comparison of environmental risks with the overall benefits of a veterinary medicine (i.e., risk management).

The draft EU–VICH–TGD was released for public consultation in the European Union in January 2006, with finalization in 2007.

3.3 REFINEMENT OF VETERINARY MEDICINAL PRODUCT (VMP) RISK ASSESSMENTS

For discussion of specific elements of effect and exposure assessments, the reader is referred to the different detailed chapters in this book. Here we discuss some specific elements that are worthy of attention when performing risk assessment for a veterinary medicine:

- Use of metabolism data in the risk assessment: the total residue approach and how to refine this
- Refinement of risk assessment based on degradation data
- Assessment of fixed combination products
- Probabilistic risk assessment of veterinary medicines

3.3.1 Metabolism and Degradation

Unlike products that may be introduced directly into the environment, such as industrial chemicals, biocides, and pesticides, veterinary medicinal products are, in most cases, metabolized by animals (and may also be degraded in manure during storage time) before their introduction to the environment (exceptions are some aquaculture and ectoparasiticidal products). Thus, in addition to the medicine itself, its metabolites may enter and could affect the environment. Although most environmental impact assessments are based on the fate and effect properties of only the parent medicine, environmental behavior of relevant metabolites should also be taken into consideration to predict if they would contribute to an increased overall risk to the environment.

With the exception of pro-drugs, the metabolites or degradation products formed generally have lower pharmacological potencies than the parent molecule and are probably also less toxic to organisms in natural ecosystems. As a result of this, VICH phase I and phase II environmental impact assessment guidelines (GL6 and GL38; VICH 2002, 2004) suggested that an assessment should be performed on the parent compound (total residue approach) in order to assess conservatively the overall environmental risk of the metabolites, on the assumption that metabolites are as toxic as the parent compound. Currently, environmental fate and effects data for metabolites of veterinary drugs are very limited.

Metabolites formed from parent veterinary medicines are generally more polar and water-soluble than the parent compound and may thus have a greater potential to run off into surface water or leach into groundwater. The degradability of metabolites, and thus their persistence, may also be significantly different from that of the parent molecule. Differences in water solubility and degradation mean that the total residue approach may not accurately predict exposures and effects, or the resulting environmental impact. False negatives (incorrectly finding no effect) are most likely when metabolites are more toxic, more mobile, or more persistent than the parent compound. Therefore, when a greater environmental risk is identified for the metabolites, further evaluation should be considered to address the specific concerns that they might cause. The VICH guidelines have briefly addressed the investigation of metabolites and stated that the data generated at phase II will be on the parent compound, but the risk assessment should also consider relevant metabolites. The relevant metabolites were defined as the excreted metabolites that represent 10% or more of the administered dose and do not form part of biochemical pathways. Thus, all metabolites formed at less than 10% of the applied dose do not normally undergo any testing, but are added to the active substance when calculating the PEC.

When evaluating the metabolites or degradation products, their overall combined impact on exposure and effect (i.e., taking into consideration both the toxicity and the amounts) should be compared to that of the parent compound. If the combined impact is still less than that from the parent molecule, it should be sufficient to perform the assessment using the total residue approach as outlined in VICH environmental impact assessment guidelines.

In some cases, risk assessment of metabolites may indicate that overall risk is reduced. For example, if the metabolite is 3 times more toxic, but only 20% is formed, its overall risk is still less than that of the parent drug molecule. A more mobile metabolite might have a concentration 20 times higher in the aquatic environment, but be 100 times less toxic to aquatic species, and have a reduced risk. However, if the reduction in toxicity is much less or the metabolite is even more toxic than the parent compound, then this may indicate a more serious risk.

Consideration of metabolites during risk assessment requires that the risk assessor understands the information obtained during ADME and residue studies. These 2 types of studies provide different windows into the understanding of metabolism and excretion due to differences in measurement techniques and animal physiology. Any observed differences in the results of these analyses could be due to the following reasons:

- The rate of metabolism for confined animals in ADME studies, which may differ from those under free field conditions in residue studies.
- Nonequivalent analytical techniques: radioactivity measurement, commonly used in ADME studies, may produce different results from chemical analysis, especially if only total residues are measured rather than individual chemical substances. Liquid chromatography tandem mass spectrometry (LC-MS-MS) analysis may produce somewhat different results than radiochromatography.
- Different types of animal feed and diets could be used in the various studies.

The environmental testing of metabolites is generally very costly, is technically challenging, and is sometimes simply impossible to perform. In order to allow for a more targeted metabolite assessment, several technical problems need to be resolved:

- The metabolites are often less stable than the parent compound and therefore present greater technical challenges in fate and effects testing.
- Obtaining a large quantity of the metabolite test substance is often hard because synthesis is difficult, as it must produce a product that has been formed by biological processes (e.g., enzymatic reactions or microbial degradations).
- Characterization of the metabolite test substance according to good laboratory practice (GLP) is not easy due to lack of appropriate analytical standards.
- Additional analytical method development and validation may be needed for the metabolites or degradates.

Alternatively, quantitative structure-activity relationships (QSARs) and quantitative structure property relationships (QSPRs) could be very useful tools to help understand the environmental and toxicological behaviors of the metabolites and degradates. In recent years, many QSAR and QSPR tools have been developed to predict the chemical properties (fate and behavior, such as mobility and persistence potential) and biological activities (effect, such as toxicity potential) of chemical molecules. However, the risk assessor should exercise caution when selecting one of these models to ensure that it suits the purpose of environmental risk assessment for veterinary medicines. For example, it would be better to employ QSAR models developed specifically for predicting toxicity behavior rather than ones for predicting drug efficacies. Similarly, if one is available, it is better to use a model developed for drug products rather than one for industrial chemicals.

In addition to using QSAR or QSPR software tools, a significant amount of preliminary toxicity and safety information on many analogs of the drug product is already available during the discovery and predevelopment stages of a drug development program. Some of these analogs might be the same metabolites and degradates of the final drug product or surrogates of the metabolites and degradates. This information can also be very useful in predicting the environmental behavior of the specific metabolites and degradates of concern.

These alternative prediction methods can play important roles in environmental impact assessment of the metabolites and degradates, as they are quick, are inexpensive, and may be easily implemented.

3.3.2 Combination Products

When a product contains more than 1 active ingredient, it might be relevant to base the risk assessment not only on the individual compounds but also on their combination(s), especially when the compounds share the same mode of action. In such cases, the sum of the PECs of these active ingredients should be compared to the trigger value in phase I in order to decide whether a phase II assessment is necessary.

A tool for the risk assessment of chemical mixtures is the prediction of their toxicities from the effects of the individual components. For that purpose, concentration addition is usually regarded as valid for mixtures of similarly acting chemicals. Whether this concept or the competing notion of independent action is more appropriate for mixtures of dissimilarly acting chemicals is still in some dispute (Backhaus et al. 2003; Junghans et al. 2006).

3.3.3 REFINEMENT OF ENVIRONMENTAL EXPOSURE PREDICTIONS

As a starting point for veterinary medicine risk assessment, the VICH guideline recommends basing the $PEC_{soil\text{-}initial}$ on the total residue approach and comparing this with the PNEC derived from a base set of toxicity tests. If the risk quotient (PEC/PNEC) is greater than 1, the PEC can be refined by taking into account degradation in the different compartments (e.g., manure or soil).

However, for the soil compartment it may be difficult to refine the PEC based on a time-weighted average. Unlike aquatic toxicity studies, the NOEC derived in soil studies is usually based on nominal concentrations, and little or no information is typically available on the fate of the substance in the medium tested. Consequently, it can only be assumed that the degradation rate of a veterinary medicine in soil after manure application equals the degradation rate found in toxicity tests. It is therefore only possible to compare the PNEC based on nominal concentrations to the initial concentrations, unless information on the fate of the medicine in the medium tested is available to calculate a time-weighted average, or if it can be anticipated that degradation will not occur in a specific test medium. This might be the case for artificial soil used in earthworm toxicity tests.

3.3.4 PROBABILISTIC RISK ASSESSMENT OF VETERINARY MEDICINES

Refinement of risk at higher tiers of risk assessment frameworks, such as those described in VICH guidance, usually involves a reduction in the conservatism of assumptions and an increase in realism, although single point estimates for deterministic estimation of PECs and PNECs remain the norm. Sometimes increased realism may be achieved through the use of more realistic models of the environment, such as estimation of a community NOEC from a mesocosm study. Alternatively, the variability and uncertainty of both exposure and toxicity data might be used to express likely environmental effects more realistically as a frequency distribution (Crane et al. 1999). Inputs to such a probabilistic risk assessment (PRA) might include comparison of a frequency distribution of modeled or measured exposure concentrations with modeled species sensitivity distributions for many species, or dose–response and population data for a single species (Posthuma et al. 2002).

Advantages of PRA are that it uses all of the available data and allows uncertainty and variability to be separated transparently in a more sophisticated characterization of risk. Disadvantages are that PRA can be data hungry and that the greater sophistication of its outputs when compared with those of deterministic approaches can make it more difficult to identify a clear risk management decision. Guidance is available on how to perform and interpret PRA (Burmaster and Wilson 1996; USEPA 1997, 1999; Warren-Hicks and Moore 1998; Posthuma et al. 2002).

PRA approaches are likely to be of most use at the highest risk assessment tiers for veterinary medicines when all lower tiers have failed. If combinations of realistic worst-case exposure and effects assessments still suggest a risk at higher tiers, PRA can help quantify risks so that decision makers base their decisions on as much information as possible. This is because PRA helps in examining all known scenarios rapidly, identifies the variables that most affect a risk forecast, and exposes the extent of uncertainty in the model, allowing improved communication of risk.

3.3.4.1 Case Study of a Probabilistic Risk Assessment for Dung Fauna

Veterinary parasiticides are widely used to treat different classes of endo- and ectoparasites of livestock. The use of these products may result in dung that contains residues of the active ingredient or metabolites that are highly toxic to different dung-related arthropod taxa, such as dung flies and dung beetles. Negative effects on the arthropod dung fauna have been detected after the use of several veterinary medicines containing different active ingredients. Consequently, this aspect has been incorporated in VICH GL38 (2004) guidance in order to protect the dung fauna and pasture function. For parasiticides intended to treat livestock reared on pasture, both dung fly larvae and dung beetle larvae studies are requested in phase II tier A. In a deterministic approach, the endpoints of these acute studies (EC50) are used with an assessment factor of 100 to derive the PNEC. This worst case is considered to be conservative enough to ensure the survival of all nontarget arthropods associated with dung (although it should be noted that the dung fly may be the target species for some ectoparasiticides). In a deterministic risk assessment, the PNEC is compared with the PEC in dung (based on the individual dosage, the number of treatments, the body weight of the animal, the mass of produced dung, and excretion events per day). The maximum concentration of the active substance in dung is estimated by taking into account the highest fraction of the dose excreted in dung in a single day.

If the resulting PEC:PNEC ratio exceeds the trigger value of 1, a risk to the dung fauna is identified. To resolve this, the PEC_{dung} should then be refined based on ADME studies of the excretion pattern and metabolism of the compound, in order to derive a reasonable maximum concentration in dung. Formation of metabolites would reduce the amount of parent compound and could be excreted via urine rather than dung. Taking a conservative approach, the refined PEC_{dung} is therefore based on the highest fraction of the administered dose excreted in dung in 1 day. Following the refinement, the PEC:PNEC ratio is recalculated and compared with the trigger value of 1. If the trigger value is still exceeded, no further recommendations are provided in VICH GL38 (2004) with respect to subsequent phase II tier B.

At this tier, a probabilistic risk assessment may usefully replace the deterministic one. A probabilistic risk assessment is more consistent with the goal of species protection at the population level rather than the survival of all individual dung beetles at all times. In order to perform a full probabilistic assessment at the population level, the following data are needed:

- Ecology and life history strategies of dung insects and their seasonal distribution
- Usage of the veterinary medicine
- Metabolism of the active ingredient
- Degradation of the parent compound and metabolites
- Effects (lethal and sublethal) of parent compound and metabolites

In the following case study, these data are provided for a theoretical ectoparasiticide licensed for the control of ticks in grazing cattle. The ectoparasiticide has to be applied topically as a pour-on to all individuals of the herd and exerts its activity for 6 weeks following each application. As the main seasonal activity of ticks is limited to the spring and early summer months, the medicine is applied 3 times (at the end of March, mid-May, and the beginning of July) in order to provide full protection over the complete pasture season (lasting from March to October). Results of laboratory studies revealed that the active compound is highly toxic to different life stages of the dung-dwelling beetle *Aphodius* spp., resulting in the death of 100% of all exposed individuals. Metabolism and excretion studies indicate that dung containing toxic amounts of the compound is excreted over a period of approximately 30 days following each administration. The combined main seasonal activity of all *Aphodius* spp. life stages lasts from April until the end of August (approximately 150 days).

Complex models that require a large number of input data have been developed for assessing the impact of parasiticides on populations of the dung fauna. However, for the purpose of a phase II tier B assessment, a simpler screening-level model may be useful for providing a worst-case assessment of impacts on the population, based on a limited data set. Such a model has been developed by Boxall et al. (2007). This modeling approach consists of the following steps (Figure 3.3):

1) Broad determination of when the sensitive stage(s) of the organism are likely to occur in dung (T), in this case mid-April through the end of August.
2) Identification of the periods when dung from animals treated with a parasiticide is toxicologically active (*t1*, *t2*, and *t3*).

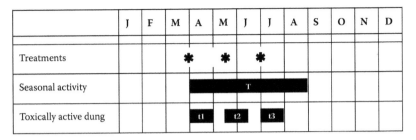

	J	F	M	A	M	J	J	A	S	O	N	D
Treatments			✳		✳		✳					
Seasonal activity							T					
Toxically active dung					t1		t2		t3				

FIGURE 3.3 Temporal distribution of main seasonal activity of *Aphodius* spp., treatment, and availability of toxically active dung.

3) Estimation of the proportion of time over which the sensitive stage(s) of the organism are theoretically capable of coming into contact with dung containing residues (q). In this case, $q = (t1 + t2 + t3) / T$. The assumptions are as follows:
  - Toxicity does not change over the period of time that dung is attractive to colonizing dung fauna.
  - The sensitive life stage (e.g., larva) does not move between dung pats.
  - The insect develops relatively quickly, so that dung colonized in periods between treatments is capable of supporting the life stage.
  - The temporal distribution of density is constant over the period that life stages colonize dung, and the distribution is independent of the prior use of veterinary parasiticides that season.

4) An estimate of the impact of the parasiticide (effectively, the percentage of individuals killed as a consequence of its use) is as follows:

$$\text{impact} = 100(p \times q \times v) \tag{3.1}$$

where:
  p = proportion of N cattle treated at any one time; and
  v = proportion of the life stage that are killed as a consequence of exposure to the highest field concentration in dung over the entire duration of this life stage.

The data in Table 3.4 were use to parameterize this simple model.
Use of Equation 3.1 leads to the following deterministic estimate of impact:

$$\text{Impact} = 100 \times 1 \times [(30 + 30 + 30)/150)] \times 1 = 60\%$$

However, data may be available on the distributions of at least some of the parameters in Table 3.4. We might assume that p remains at 1, as this makes veterinary sense for animal health. In contrast, data may be available to show that v varies uniformly from 0.6 to 1, T may vary uniformly from 120 to 170 days, and $t1$–$t3$ may vary logarithmically because of degradation, with a lower 5th percentile

TABLE 3.4

Parameters for estimating parasiticide impacts on dung insect populations

Parameter	Value
Proportion of cattle treated on each occasion (p)	1
Proportion of time-sensitive dung beetle stages in contact with dung (q)	0.7
Proportion of life stage killed (v)	1
Duration over which exposure could occur (days, T)	150
Duration of exposure 1 (days, t1)	30
Duration of exposure 2 (days, t2)	30
Duration of exposure 3 (days, t3)	30

of 5 and an upper 95th percentile of 15. The results from running 10000 Monte Carlo simulations using these distributions in Crystal Ball software (Oracle Crystal Ball, Denver, CO; http://www.crystalball.com) are shown in Figure 3.4. The resulting forecast distribution shows that more than 99% of values lie below the original deterministic value of 60% of dung insects killed.

Another valuable output from a probabilistic analysis is a sensitivity analysis that shows which model components contribute most to the final outputs. In this example, $t1$ to $t3$ contributed most (~25% each) to model sensitivity, with v contributing 17.2% and T contributing 7.5%. This means that it may be worth investing further resources in characterizing $t1$ to $t3$ more accurately to provide more accurate estimates of effect.

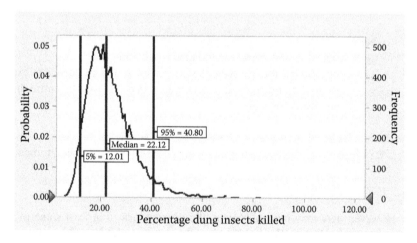

FIGURE 3.4 Distribution of effect values in a simple probabilistic model of dung insect toxicity.

Use of probabilistic risk assessment almost inevitably leads to a more explicit requirement to define protection goals and acceptability criteria because it does not produce simple pass–fail threshold outputs. The following suggestions are based on the premise that the output from a higher tier assessment is a probabilistically derived estimate of the likelihood of adverse effects. The output should have the following characteristics:

1) The intensity of the effect and its variation is known or predicted with respect to the following:

  • Types of environments or habitats affected,
  • Temporality, and
  • Spatial extent of the effects.

2) The effects are segregated by class of organism and the function of these organisms in the ecosystem, and their recovery potential is known or can be predicted.
3) The likelihood of affecting specially protected areas (nature reserves) is known or can be predicted.
4) The likelihood of affecting socially important (endangered) organisms is known or can be predicted.

These data can then be used to classify and apportion the effects into the types of categories shown in Table 3.5.

How these categories of effect are used in the decision-making process would be dependent on the benefits and other social and economic considerations. These would vary from one case to another but would also need to be described so as to ensure the transparency of the decision-making process.

In the hypothetical dung insect example above, the deterministic estimate of effects was 60%, which is class 4 in Table 3.5. In contrast, even the 95th percentile of the probabilistic distribution would place the results in class 3. Such a difference could affect decisions about product authorization or risk mitigation requirements.

3.4   RISK MANAGEMENT

Risk mitigation is an essential part of the evaluation of potential products and the management of field contamination. Most measures are aimed at reducing exposure to veterinary medicines, starting with the selection of potential products by a company because toxicity to nontarget organisms and persistence in the environment need to be considered in the early stages of product development. When the potential product is under review for approval or authorization, the risk assessor can stipulate the use of mitigation measures to restrict the risk associated with a product to an acceptable level. After authorization or approval (i.e., during use, disposal, or cleanup of spills), knowledgeable consumers or site managers can implement mitigation measures.

TABLE 3.5

Criteria for classifying known or predicted effects of veterinary medicines in the ecosystem

Class	Criterion	Description
1	Effect not likely	No statistically significant effects (< 5% probability of any responses) known or predicted as a result of the use of the VMP.
2	Slight effect	Known or predicted effects slight or transient (> 5% < 20% probability of occurrence either spatially or temporally), with recovery occurring within 2 to 3 generations of the affected organisms or in less than a season (until the following spring or normal use period of veterinary medicines for disease/parasite management purposes).
3	Pronounced but restricted short-term effect	Known or predicted effects pronounced but transient (> 20% < 50% probability of occurrence either spatially or temporally), with recovery occurring within 2 to 3 generations of the affected organisms or in less than a season.
4	Pronounced and widespread short-term effect	Known or predicted effects pronounced but transient (> 50% probability of occurrence either spatially or temporally), with recovery occurring within 2 to 3 generations of the affected organisms or in less than a season.
5	Pronounced long-term effect	Known or predicted effects pronounced (> 50% probability of occurrence either spatially or temporally), with recovery occurring in more than 1 season.

[a] Effects may be structural, functional, or aesthetic, depending on the protection goals.

3.4.1 RISK MITIGATION MEASURES WITHIN PRODUCT AUTHORIZATION OR APPROVAL

Risk mitigation measures are often required by the regulator to reduce the risk from use of a veterinary medicine during the approval or authorization process. The efficacy of such measures should be substantiated by data in the dossier. When removing an indication or even target animal from the label, such a proposal may obviate the need for further testing.

To be effective, risk mitigation measures should meet the following criteria. They should

- reduce environmental exposure and transport of the veterinary medicine;
- be feasible with respect to agricultural practice (i.e., be likely to be followed in practice);
- be consistent with applicable regulations; and
- have scientifically demonstrable effects.

A wide variety of strategies have been employed to reduce environmental exposures, and such risk mitigation measures are normally communicated on the product label. A label might specify, for instance, that treated animals should not have direct access to surface water and ditches, or access should be restricted for a period of time. In order to protect sensitive areas, the label may require a buffer zone (i.e., a strip of land) between the application site and surface waters. Labels for persistent products may restrict repeated use in the same location, or the frequency of use. Restricting repeated use can also be effective in places where repeated exposure is likely to lead to declines in nontarget populations. In cases in which local regulations may apply, such as for disposal, the wording may refer the reader to other guidance. For instance, in the European Union, standard advice for disposal for unused veterinary medicines reads, "Unused medicines should be disposed of in accordance with national requirements."

The storage and handling of veterinary medicines in manure from treated animals are a special concern because the medicine can be leached into the environment and because manure fauna and animals that might feed on manure fauna are also potentially at risk. Risk mitigation measures for contaminated manure could specify storing the manure for a period of time, adding substances that will reduce the hazard of the medicine, or restricting the frequency or rate of application of manure onto fields. For instance, one label for pigs requires manure from treated pigs to be stored for 3 months prior to spreading and incorporating into land. For highly mobile and persistent veterinary medicines, labels can restrict the application rate at groundwater-sensitive sites. Labeling for aquaculture can similarly specify disposal options. Some aquaculture products require a period of time in a settling basin or addition of a detoxifying agent. Additional measures may be required to ensure the safety of animals that may be used for human food, although they are not presented here because human health concerns are beyond the scope of this book.

Risks due to veterinary medicines are not the only reason that the spreading of manure is restricted. Many jurisdictions restrict this procedure because of concerns about nutrient inputs to surface waters. When spreading manure, buffer zones of 10 meters between application site and surface water may be recommended in good agricultural practice guidelines. In some EU countries, these buffer zones are included in the legislation regulating manure spreading. However, it should not be assumed that requirements imposed in order to control nutrients are sufficient to control all other environmental risks arising from the use of veterinary medicines.

Communication to the individuals responsible for carrying out the mitigation measure is often a significant challenge. An extensive communication strategy is needed to ensure that individuals are aware of their label responsibilities. Mitigation measures should be based on a realistic understanding of these communication challenges, including the background knowledge of the responsible individuals. Information on a label that is read only by a veterinarian may not be communicated to the person who purchases and spreads manure from a variety of sources. Requiring a "withdrawal period" (e.g., storing the manure for 3 months before spreading) may not be realistic if that knowledge must be passed on to future owners of the manure.

Another reason why reliance on risk mitigation measures is somewhat risky is that enforcement of label requirements is often lacking. Although the approval or authorization process is regulated at the national level (or, in the case of the European Union, a centralized procedure at the EU level), the enforcement of risk mitigation measures is usually at a lower level. Risk assessors should have a realistic awareness of the patchwork of enforcement laws, procedures, and capabilities that will control the implementation of the measures. For example, with the exception of maximum residue level (MRL) regulations for food-producing animals, there is no requirement for surveillance at the local level in the European Union.

In some cases, a market authorization may be granted as long as stipulated data are provided by a certain time. The product is then placed on the market although the assessment has not yet been finalized. Within a defined period the applicant has to provide additional data. After the data have been provided the assessment is completed, and a decision is made on whether the market authorization is extended or not. These types of exemptions have been issued to products with a high therapeutic use. Market exemptions are also granted when fate- and effects-monitoring data needed for assessment require veterinary use on a larger scale (e.g., for fish medicines).

3.4.2 RISK ASSESSMENT AND MANAGEMENT BEYOND AUTHORIZATION OR APPROVAL

3.4.2.1 Communication Challenge

Individual perception of risk is influenced by a variety of factors. Lack of personal control over and dread of the potential or real consequences of the risk, conflict between experts, uncertainty, and unfair distribution of risks and benefits will all contribute to a heightened perception of the risk. Effective risk communication must take these factors into account.

Research in the social sciences has turned up complex relationships between scientific results and assessment, trust, and public perception (Douglas 2000). The public's perception of risks may well diverge significantly from that of specialists (Hansen et al. 2003; Frewer 2004) because an individual's perception of risk depends upon an often intuitive and emotional judgment of the probability of occurrence and the severity of the consequences of that risk. This perception is usually a judgment that is made without consideration of associated benefits, and risks only become acceptable to an individual when they are able to balance them with these benefits. However, even if individuals agree on the degree of risk, they may still disagree on its acceptability because of differences in their level of expertise and education, their gender, or their personal values (Tait 2001a; Frewer 2003; Frewer et al. 2004).

When the public is questioned in opinion polls, concerns about chemicals are greatest on issues of human health and food quality (Dunlap and Beus 1992; Anon 2000; Crane et al. 2006). However, potential environmental effects are also an issue for a substantial number of people, particularly if attractive species could be affected (Crane et al. 2006). Despite these views, Tait et al. (2001) found from a review of the literature on chemicals and public values that public attitudes and values, on one hand, and actual behavior, on the other, are only weakly correlated, often because of intervening variables such as price (e.g., of organic food versus food grown with the use of pesticides or veterinary medicines).

Experts and the public tend to rank the relative generic risks from chemical contaminants consistently. For example, Slovic (2000) asked experts and laypeople to rank the perceived risks of 30 potentially hazardous activities. In his survey, the laypeople ranked "Pesticides" ninth, whereas the experts ranked them eighth. This convergence is in marked contrast to activities associated with significant public dread (Perrow 1999), such as "nuclear power," which was ranked as the most important hazard by laypeople but was ranked only 20th by experts.

Many scientists and industrialists believe that greater public understanding of science is the solution to public attitudes that seem to be irrational or are at variance with expert views or the actual behavior of the public. However, social science studies show that this is unlikely to be a complete solution because once a person's mind is made up about fundamental values, he or she will normally use only the scientific information that supports his or her position, ignoring the science that does not (Tait 2001b).

The completion of the authorization or approval process is not the end of opportunities to mitigate risk. Labels are only one means of educating the agricultural community about managing risks from veterinary medicines. Development of new farming techniques for the agricultural community may improve the mitigation of risk. For example, some South African farmers have automated dosing regimes for their cattle, resulting in reductions in the amounts of ectoparasiticidal products that are needed. Farmers who graze their cattle on protected lands may be given additional information from the authorities to mitigate risk on those lands. Consumers can also help to mitigate the risks of contamination. They can select products that have fewer nontarget effects, less waste, more effective disposal options, or less persistence. The ability to make these types of decisions demands a well-educated and environmentally concerned consumer.

Special efforts at communication may be necessary in order to reach the appropriate audiences. "Green labeling" of products is one way that this information can be effectively communicated. Green labels can contain language that specifies the safety to particular faunal groups, such as bats, based on test information submitted as part of the veterinary medicine regulatory dossier. Such language may offer competitive advantages to those products. For example, the Poison Working Group of the Endangered Wildlife Trust has worked to establish an education and "green branding" program to increase populations of red-billed and yellow-billed oxpeckers, which feed on ticks infesting game and livestock.

Oxpecker population and range have been significantly reduced by the use of arsenic-based ectoparasiticides. In order to reduce the poisoning of these threatened birds, the Endangered Wildlife Trust established the Poison Working Group to promote environmentally responsible management of external parasites on livestock in southern Africa. The Poison Working Group compiled and distribute a bilingual oxpecker compatibility chart to educate farmers about the safety of tick control products on the market and negotiated with the National Department of Agriculture to allow manufacturers of "oxpecker-compatible" dips to transition to "green brand" products that are safe for the oxpecker. They monitor the increase in the number of farmers who use these products and supplement natural reintroduction of oxpeckers in areas where only oxpecker-compatible products are used. They are also documenting the economics of biological control provided by oxpeckers.

In contrast to positive green labeling, negative labeling such as "Product has not been assessed for ecological risks" could be required.

From the discussion above, it is clear that the audience to whom the risks of veterinary medicines need to be communicated includes several groups. Risk managers and other professional risk specialists, political decision makers, food retailers, farmers, veterinarians, nongovernmental organizations (NGOs), and the public all need tailored risk communication approaches:

1) Risk managers and specialists will require clear, quantitative communication of the environmental risks of a veterinary medicine from a risk assessor, with supporting data to justify any conclusions.

2) Political decision makers and food retailers will also require clear communication of the environmental risks, but are unlikely to want more than a qualitative assessment from trusted experts, and they will also need reassurance that any identified risks will be acceptable to the wider public.

3) Farmers and veterinarians are the stakeholders most likely to consider that the animal health benefits of veterinary medicines outweigh environmental risks. Communication of any risks, the means to control them, and the consequences to the individual if these are ignored should therefore be made very explicit on packaging and during face-to-face discussions. For example, the Environment Agency of England and Wales has greatest success in influencing behavior when site visits to educate potential dischargers, including farmers, are combined with a clearly articulated policy to prosecute those who knowingly pollute the environment.

4) The public is most likely to trust information on environmental risks if these risks are reasonably well understood, experts have reached consensus, environmental risks are placed in the context of animal health benefits and are outweighed by them, and risks to human health can be excluded.

5) Green nongovernmental organizations are likely to trust information on the same basis as the public, with one exception. If the policy of the NGO

is to reject the use of a medicine on a matter of principle (e.g., opposition to intensive rearing practices that involve use of that medicine), then it is likely that evidence of adverse environmental effects will be used to support this position, irrespective of any balancing benefits.

3.4.2.2   Incidence Reporting and Pharmacovigilance

After the marketing authorization of a veterinary medicine, the marketing authorization holder (MAH) is obliged to collect information on adverse events related to the product and communicate this to the relevant competent authorities. Different types of adverse events fall under the scope of such "pharmacovigilance," namely, clinical safety (adverse events in the treated animal), extralabel use (the product has not been used in accordance to the data sheet, e.g., incorrect dosage, nonlicensed target species, and contraindicated diseases), lack of expected efficacy (the product has not done what it claims to do although it was administered correctly), MRL violations (approved residue limits in food animals are violated), human exposure, and environmental problems. It is the obligation of the MAH to establish a thorough pharmacovigilance system that guarantees the collection and collation of adverse events related to any veterinary medicine. Pharmacovigilance is therefore in principle a well-established tool to assess the environmental safety of a veterinary medicine following marketing authorization. Pharmacovigilance information is a major input to reevaluation of the risk–benefit balance of a medicine in the course of marketing authorization renewals. Furthermore, the pharmacovigilance data collected by the MAH have to be submitted at regular intervals to the competent authorities, enabling a continuous assessment of the medicine. If serious risks are identified within this pharmacovigilance process, a marketing authorization may be suspended or not renewed.

The United Kingdom provides a typical case study of an incident-reporting scheme. The reporting of environmental incidents involving veterinary medicines became part of the UK Suspected Adverse Reaction Surveillance Scheme (SARSS) in 1998. The Veterinary Medicines Directorate is the licensing authority for veterinary medicines in the United Kingdom and runs the SARSS nationally. The SARSS runs parallel to the pharmacovigilance scheme. Although the MAH are legally required to provide pharmacovigilance data relating to environmental incidents, it is recognized that they are unlikely to be the source of many of these reports, as most pollution incidents, especially of surface waters, will be notified directly to the environment agencies either by their employees during routine monitoring work or by the public if they notice fish kills or other extreme events. Incidents are therefore generally reported by the Environment Agency of England and Wales, and the Scottish Environment Protection Agency. This provides a framework for subsequent investigation and reporting of these incidents: the agencies can respond to the incidents rapidly (depending on the severity of the reported incidents) and collect additional information such as environmental samples for chemical and biological analyses. The Wildlife Incident Investigation Scheme (WIIS, for animal poisoning) also reports to the SARSS team.

However, useful feedback from pharmacovigilance may be weak. Pharmaco-vigilance and other incident-reporting schemes can usually identify only gross examples of impacts, and reliance on them has the following problems.

3.4.2.2.1 Reporting of an Adverse Event

The collection of pharmacovigilance data by the MAH is a passive process, totally dependent on the reporting of adverse events by veterinarians, pharmacists, or the animal owners (these reporters can contact either the MAH or a competent authority that will subsequently inform the MAH). As is commonly recognized, only a fraction of all adverse events occurring after the use of a VMP are in fact reported to the competent authority or the MAH. The reporting is on a voluntary basis, and none of the above-mentioned parties is obliged to report an adverse event. In fact, there may be a reluctance to report due to a lack of general aware-ness about pharmacovigilance schemes, an increased workload, or the consider-ation of an adverse event as nonrelevant. In the case of environmental problems, an adverse event might be related to off-label use of the product (e.g., improper disposal or overdosage). In addition, the user might also presume that he or she could become liable for any environmental damage. In both cases, the user will most likely not be willing to come forward with adverse event information.

3.4.2.2.2 Validity of an Adverse Event

Four minimum criteria have to be fulfilled to make a report into a full valid adverse event case. The reporting source has to be identifiable, details of the treated animal(s) have to be known, an adverse reaction has to be described, and a veterinary medicine has to have been given. Although the first 3 criteria might be easily available in most cases, the link of an environmental problem to a specific veterinary medicine (1 single product with a brand name, not simply a product group or an active ingredient) might be difficult, especially when the environmental problem has been identified by a person not directly involved in the use of the product. In such a case, the adverse reaction will not be included in the pharmacovigilance system by the MAH and consequently not be reported to the competent authorities.

3.4.2.2.3 Identifying Environmental Problems

In reality, there are probably only a few environmental problems that can be iden-tified by a user and then be linked to the use of the product (e.g., crayfish kills after sheep dip exposure, as reported in the United Kingdom in relation to the use of particular sheep dip products). In contrast, most negative effects on the envi-ronment will not be obvious to the user, such as toxicity toward *Daphnia* or influ-ences on earthworm populations or microbial activity.

The UK SARSS scheme has identified the following issues when identifying environmental incidents:

- Veterinary medicines are not actively monitored in the environment by the UK environment agencies, with the exception of sheep dip chemicals in controlled surface waters. As a result, these comprise the great majority of environmental SARSS reported to the VMD.

- Veterinary medicines are not routinely monitored in soil and groundwater.
- The scheme is reliant on reactive investigation stimulated by reports from the public or problems identified during infrequent routine monitoring. This will result in an underestimation of the number of incidents. Indeed, there is evidence of considerable underreporting of even sheep dip environmental incidents in surface waters. Focused studies in the Environment Agency's Welsh region and in Scotland have shown that active strategic monitoring of areas in which sheep are dipped reveals a greater frequency and severity of incidents than does reactive investigation.
- The WIIS also suffers from being a reactive scheme that depends upon members of the public noticing an adverse environmental incident. This makes it more likely that a report will be made if, say, a single red kite dies from misuse of a veterinary medicine than if there is widespread leaching of an antimicrobial medicine into groundwater.

3.4.2.2.4   Incidence Calculation

After marketing authorization of a VMP, pharmacovigilance data are compiled at regular intervals in Periodic Safety Update Reports (European Union) or Annual Reports (United States). These documents are used to consider if the risk:benefit ratio has been altered. For this purpose the sales volume of the product (expressed as administered doses) is compared with the affected individuals reported in the adverse events. However, in common practice, incidences are calculated for cases of clinical safety, extralabel use, and MRL violation only. Consequently, any changes in the risk:benefit ratio due to environmental problems will not be taken into account. Even if incidence calculations were performed for environmental problems, it is likely that there would be little resulting concern about risks and benefits because of the artificially limited number of adverse events that are reported, for the reasons cited above.

Environmental risk assessment is unlike human or target species risk assessment because of the much wider range of species and exposure pathways that must be considered. This makes accurate prospective risk assessment difficult at the authorization stage. Therefore, a regulatory scheme that does not involve credible postauthorization monitoring is likely to suffer from an unknown number of false negatives, in which the environmental risks of chemicals are underestimated. There is a need therefore for more active strategic monitoring of the environmental fate and effects of those veterinary medicines that have the potential to cause harm to the environment.

3.4.3   Retrospective Risk Assessment

As described in the preceding sections, prospective risk assessments aim to predict the potential effects resulting from the use and release of a particular chemical, generally within the premarket approval process for a specific product containing this chemical. Retrospective risk assessments, although they may follow similar protection goals, differ in their approach. They may try to identify the impacts resulting from a particular source of chemicals (e.g., industrial effluents), attempt to establish the cause(s) of measured ecological effects (e.g., decrease in invertebrate populations), or assess both the causes and effects of measured exposures to particular chemical(s) (e.g., sheep dips). Their scale also varies from regional to site-specific assessments.

The challenges when conducting retrospective assessments lie in the choice of endpoints, the interpretation of effects, and the identification of causes. Endpoints must be measurable, ecologically relevant, susceptible to stressors, and relevant to protection goals. However, measuring endpoints is in itself insufficient to indicate whether the system has been impacted. Natural systems are complex and variable assemblages of species and conditions; the reality of effects apparently measured and their significance are often debated. Results must be interpreted by comparison with a reference state. This may be an "acceptable" state (e.g., in the European Union's Water Framework Directive, "good status" means a slight deviation from "high status," which is the best status achievable). In the absence of such a standard, a base state may be used, for example proportional changes are measured and the direction of change is judged as an improvement or decline. Even when impacts have been identified, it is often difficult to assign causation with the degree of certainty required, particularly if indirect effects (e.g., via the food web) are suspected or when multiple causes can be proposed (e.g., complex mixtures of chemicals associated with changes in habitat).

When identifying terrestrial and aquatic sites for retrospective ecological risk assessments of veterinary medicines, arid and semiarid regions may present unique case studies for exposure and effects assessments. Aquatic systems in more arid environments generally receive limited or no upstream dilution of contaminant inflows; appreciable dilution occurs only following seasonal storm events (Brooks et al. 2004). Because transport of veterinary medicines and other contaminants such as nutrients, pathogens, and sediments via runoff from concentrated animal-feeding operation (CAFO) lagoons and application fields has been demonstrated (McFarland and Hauck 1999; Geary and Davies 2003; Boxall et al. 2004; Orlando et al. 2004), arid and semiarid headwater streams located adjacent to CAFOs may represent worst-case scenarios for aquatic exposure and potential effects (Brooks et al. 2006).

One such example in the United States is the North Bosque River, located in the Brazos River basin of central Texas. Water quality in the North Bosque River watershed is impacted due to nonpoint source nutrient pollution from dairy CAFOs (McFarland and Hauck 1999). Runoff from these CAFOs largely occurs during stormflow events, which has resulted in elevated nutrient loads to the river, and hypereutrophication of Waco Lake, located downstream at the junction of the North, Middle, and South Bosque Rivers.

The Texas Commission on Environmental Quality (n.d.) has developed a total maximum daily load (TMDL) and implementation plan for soluble reactive phosphorus (SRP) in the watershed. The goal of the TMDL and implementation plan is to reduce the annual average concentration of instream SRP by approximately 50% (see http://www.tceq.state.tx.us).

Development of state-of-the-science fate and transport models at the watershed scale may allow for predictions of instream contaminant concentrations and exposures to veterinary medicines under various scenarios. For example, the soil water and assessment tool (SWAT; Arnold and Fohrer 2005), a physically based watershed model that incorporates landscape features (landcover, slope, and soils) through an interface with ArcGIS®, can be used to identify various contributions of nutrients and sediments to watershed water quality impairment. Because veterinary medicines have not been studied in dairy CAFOs found in the North Bosque River watershed, it remains untested whether transport of contaminants (e.g., nutrients and pathogens) that often co-occur with veterinary medicines may be used as surrogate measures for veterinary medicine introductions to this or other aquatic systems.

3.4.4   POSTMARKET MONITORING AND REMEDIATION

No risk assessment process is entirely foolproof. Monitoring can be used to validate or to understand better the risks and potential impacts from actual use, thus improving the risk, exposure, and effects assessments. However, this information may not be available through or during the premarket approval process. The standard preauthorization guidance may not require data for combination products, pathways of exposure, indirect effects, or potential effects on less studied species or environmental compartments (e.g., amphibians or sediments). Likewise, monitoring may demonstrate the effectiveness of required risk mitigation measures. Monitoring enables the risk assessor to identify potential impacts of older chemicals that may not have been fully assessed for environmental safety or new chemicals that have been misused. Monitoring can also be used to determine the effects of veterinary medicines in the context of multiple contaminants or other stressors. Likewise, monitoring data can help evaluate combined exposures to different (veterinary and nonveterinary) uses of the same chemical. Finally, monitoring may identify hotspots where chemicals have been used improperly and help risk assessors of hazardous waste sites identify the environment of interest.

The design of the studies will depend on the monitoring goals: these goals will help to determine the scope of the effort. Studies may be local, regional, or national and focus on a particular assessment or surveillance scheme. They may be short term or long term, such as to demonstrate the efficacy of a remediation effort or other management activity.

3.4.4.1   Monitoring Endpoints

Veterinary medicines released into the environment can be measured in many different ways. Appropriate monitoring strategies should consider the various pathways of exposure that are characteristic of medicines. However, interpretation of these data is not always straightforward. One of the most clear-cut interpretive tools is the use of numerical chemical standards and criteria to compare with measured environmental concentrations. In most cases, however, standards and criteria have not yet been set for veterinary medicines. Alternatively, biomarkers of toxic effects and biological indices have also been developed and can be used to assess environmental impact, especially in the aquatic arena, but these too are often difficult to interpret.

Regardless of the endpoints used, the strategies for designing monitoring programs have many commonalities. Substances for analysis must be prioritized. This informs monitoring programs, from the targeted monitoring at farm level to the identification and inclusion of substances for environmental quality standards and landscape-level programs. The design of monitoring programs should be informed by all available data. Obtaining information on usage patterns and spatial distribution is a critical first step. Unless the aims are specifically to understand pathways of exposure, monitoring data should consist of a holistic approach — chemical data, biological data, and ecosystem data (populations). These are all discussed in detail in subsequent chapters.

REFERENCES

Anon. 2000. Termites and the public: attitude toward pesticides. Pest Cont Technol 28:50–52.

Arnold JG, Fohrer N. 2005. SWAT2000: current capabilities and research opportunities in applied watershed modeling. Hydrol Process 19:563–572.

Backhaus T, Altenburger R, Arrhenius A, Blanck H, Faust M, Finizio A, Gramatica P, Grote M, Junghans M, Meyer W, Pavan M, Porsbring T, Scholze M, Todeschini R, Vighi M, Walter H, Grimme LH. 2003. The BEAM-project: prediction and assessment of mixture toxicities in the aquatic environment. Cont Shelf Res 23:1757–1769.

Boxall ABA, Fogg LA, Blackwell PA, Kay P, Pemberton EJ, Croxford A. 2004. Veterinary medicines in the environment. Rev Environ Contam Toxicol 180:1–91.

Boxall ABA, Sherratt TN, Pudner V, Pope, LJ. 2007. A screening level index for assessing the impacts of veterinary medicines on dung flies. Environ Sci Technol 41:2630–2635.

Brooks BW, Riley TM, Taylor RD. 2006. Water quality of effluent-dominated stream ecosystems: ecotoxicological, hydrological, and management considerations. Hydrobiologia 556:365–379.

Brooks BW, Stanley JK, White JC, Turner PK, Wu KB, La Point TW. 2004. Laboratory and field responses to cadmium in effluent-dominated stream mesocosms. Environ Toxicol Chem 24:464–469.

Burmaster DE, Wilson AM. 1996. An introduction to second-order random variables in human health risk assessments. Hum Ecol Risk Ass 2:892–919.

Calow P. 1998. Ecological risk assessment: risk for what? How do we decide? Ecotoxicol Environ Safety 40:15–18.

Crane M, Grosso A, Janssen C. 1999. Statistical techniques for the ecological risk assessment of chemicals in freshwaters. In: Sparkes T, editor. Statistics in ecotoxicology. Chichester (UK): John Wiley, p 247–278.

Crane M, Norton A, Leaman J, Chalak A, Bailey A, Yoxon M, Smith J, Fenlon. 2006. Acceptability of pesticide impacts on the environment: what do United Kingdom stakeholders and the public value? Pest Man Sci 62:5–19.

Defra. 2004. Explanatory memorandum to statutory instruments. The controls on nonyl-phenol and nonylphenol ethoxylate regulations 2004, SI 2004 No. 1816. Department for the Environment, Food and Rural Affairs, London, UK. www.opsi.gov.uk/si/em2004/uksiem_20041816_en.pdf.

Douglas H. 2000. Inductive risk and values in science. Philos Sci 67:559–579.

Dunlap RE, Beus CE. 1992. Understanding public concerns about pesticides: an empirical examination. J Consumer Affairs 26:418–438.

[EC] European Commission. 1990. Council Directive 90/676/EEC of 13 December 1990 amending Directive 81/851/EEC on the approximation of the laws of the Member States relating to veterinary medicinal products Official Journal *L373, 31.12.1990.*

[EC] European Commission. 1993. Council Directive 93/39/EEC of 14 June 1993 amending Directives 65/65/EEC, 75/318/EEC and 75/319/EEC in respect of medicinal products. Official Journal L214, 24/08/1993.

[EMEA] European Agency for the Evaluation of Medicinal Products. 1998. Note for Guidance: Environmental risk assessment for veterinary medicinal products other than GMO-containing and immunological products. London, UK: European Agency for the Evaluation of Medicinal Products. Rapport nr. EMEA/CVMP/055/96.

[EMEA] European Agency for the Evaluation of Medicinal Products. 2001. Discussion paper on environmental risk assessment of non-genetically modified organism (non-GMO) containing medicinal products for human use. CPMP/SWP/4447/00 draft corr. London (UK): European Agency for the Evaluation of Medicinal Products.

[EMEA] European Agency for the Evaluation of Medicinal Products. 2003. Note for guidance on environmental risk assessment of medicinal products for human use. CPMP/SWP/4447/00 draft. London (UK): European Agency for the Evaluation of Medicinal Products.

[EMEA] European Agency for the Evaluation of Medicinal Products. 2005. Guideline on the environmental risk assessment of medicinal products for human use. CHMP/SWP/ 4447/00 draft. London (UK): European Agency for the Evaluation of Medicinal Products.

European Crop Protection Association. N.d. http://www.ecpa.be.

European Parliament. 2001a. Directive 2001/82/EC of the European Parliament and of the Council of 6 November 2001 on the community code relating to veterinary medicinal products. Official Journal L 311, 28/11/2001, p 0001–0066. http://ec.europa.eu/enterprise/pharmaceuticals/eudralex/homev1.htm.

European Parliament. 2001b. Directive 2001/83/EC of the European Parliament and of the Council of 6 November 2001 on the community code relating to medicinal products for human use. Official Journal L 311, 28/11/2001, p 0067–0128. http://ec.europa.eu/enterprise/pharmaceuticals/eudralex/homev1.htm.

European Parliament. 2004a. Directive 2004/27/EC of the European Parliament and of the Council of 31 March 2004 amending Directive 2001/83/EC on the community code relating to medicinal products for human use. Official Journal L 136, 30/04/2004, p 0034–0057. http://ec.europa.eu/enterprise/pharmaceuticals/eudralex/homev1.htm

European Parliament. 2004b. Directive 2004/28/EC of the European Parliament and of the Council of 31 March 2004 amending Directive 2001/82/EC on the community code relating to veterinary medicinal products. Official Journal L 136, 30/04/2004, p 0058–0084. http://eur-lex.europa.eu/LexUriServ/LexUriServ.do?uri=OJ:L:2004:1 36:0058:0084:EN:PDF.

European Parliament. 2004c. Regulation (EC) No 726/2004 of the European Parliament and of the Council of 31 March 2004 laying down community procedures for the authorisation and supervision of medicinal products for human and veterinary use and establishing a European Medicines Agency. Official Journal L 136, 30/04/2004, p 0001–0033. http://ec.europa.eu/enterprise/pharmaceuticals/eudralex/vol-1/reg_2004_726/reg_2004_726_en.pdf.

Frewer L. 2003. Societal issues and public attitudes towards genetically modified foods. Trends Food Sci Technol 14:319–332.

Frewer L. 2004. The public and effective risk communication. Toxicol Lett 149:391–397.

Frewer L, Lassen J, Kettlitz B, Scholderer J, Beekman V, Berdal KG. 2004. Societal aspects of genetically modified foods. Food Chem Toxicol 42:1181–1193.

Geary PM, Davies CM. 2003. Bacterial source tracking and shellfish contamination in a coastal catchment. Wat Sci Technol 47:95–100.

Hansen J, Holm L, Frewer L, Robinson P, Sandoe P. 2003. Beyond the knowledge deficit: recent research into lay and expert attitudes to food risks. Appetite 41:111–121.

Junghans M, Backhaus T, Faust M, Scholze M, Grimme LH. 2006. Application and validation of approaches for the predictive hazard assessment of realistic pesticide mixtures. Aquatic Toxicol 76:93–110.

McFarland AMS, Hauck LM. 1999. Relating agricultural land uses to in-stream stormwater quality. Journal of Environmental Quality 28:836–844.

National Archives and Records Administration. 2004. Code of federal regulations. http://www.access.gpo.gov/nara/cfr/waisidx_01/21cfr25_01.html.

National Environmental Policy Act. 1969. NEPA, the National Environmental Policy Act [42 U.S.C.4321 et seq.].

Orlando EF, Kolok AS, Binzcik GA, Gates JL, Horton MK, Lambright CS, Gray LE Jr, Soto AM, Guillette L Jr. 2004. Endocrine-disrupting effects of cattle feedlot effluent on an aquatic sentinel species, the fathead minnow. Environ Health Perspect 112:353–358.

Perrow C. 1999. Normal accidents: living with high-risk technologies. Princeton (NJ): Princeton University Press.

Posthuma L, Suter GW, Traas TP, editors. 2002. Species sensitivity distributions in ecotoxicology. Boca Raton (FL): Lewis.

Slovic P. 2000. The perception of risk. London (UK): Earthscan.

Tait J. 2001a. More Faust than Frankenstein: the European debate about the precautionary principle and risk regulation for genetically modified crops. J Risk Res 4:175–189.

Tait J. 2001b. Pesticide regulation, product innovation and public attitudes. J Environ Monitor 3:64N–69N.

Tait J, Bruce A, Lyall C. 2001. Studies on people's values in relation to chemicals and their effects on humans and the natural environment: a literature review. Report to the Royal Commission on Environmental Pollution. SUPRA Paper 23. Edinburgh (Scotland): Royal Commission on Environmental Pollution.

Texas Commission on Environmental Quality. N.d. http://www.tceq.state.tx.us.

[USEPA] US Environmental Protection Agency. 1997. Guiding principles for Monte Carlo analysis. EPA/630/R-97/001, Risk Assessment Forum. Washington (DC): US Environmental Protection Agency. http://www.epa.gov/ncea/raf/montecar.pdf.

[USEPA] US Environmental Protection Agency. 1999. Report of the workshop on selecting input distributions for probabilistic assessments. EPA/630/R-98/004, Risk Assessment Forum. Washington (DC): US Environmental Protection Agency.

[VICH] International Cooperation on Harmonization of Technical Requirements for Registration of Veterinary Products. 2002. Environmental impact assessment (EIAs) for veterinary medicinal products (veterinary medicines) — phase I. VICH GL6 (ecotoxicity phase 1) June. http://vich.eudra.org.

[VICH] International Cooperation on Harmonization of Technical Requirements for Registration of Veterinary Products. 2004. Environmental impact assessment (EIAs) for veterinary medicinal products (veterinary medicines) — phase II. VICH GL38 (ecotoxicity phase II) October. http://vich.eudra.org.

[VMD] 2005. Sales of Antimicrobial Products Authorized for Use as Veterinary Medicines, Antiprotozoals, Antifungals, Growth Promoters, and Coccidiostats, in the UK in 2004. Veterinary Medicines Directorate, Addlestone, UK.

Warren-Hicks WJ, Moore DRJ. 1998. Uncertainty analysis in ecological risk assessment. Pensacola (FL): SETAC Press.

4 Exposure Assessment of Veterinary Medicines in Aquatic Systems

Chris Metcalfe, Alistair Boxall, Kathrin Fenner,
Dana Kolpin, Mark Servos, Eric Silberhorn, and
Jane Staveley

4.1 INTRODUCTION

The release of veterinary medicines into the aquatic environment may occur through direct or indirect pathways. An example of direct release is the use of medicines in aquaculture (Armstrong et al. 2005; Davies et al. 1998), where chemicals used to treat fish are added directly to water. Indirect releases, in which medicines make their way to water through transport from other matrices, include the application of animal manure to land or direct excretion of residues onto pasture land, from which the therapeutic chemicals may be transported into the aquatic environment (Jørgensen and Halling-Sørensen 2000; Boxall et al. 2003, 2004). Veterinary medicines used to treat companion animals may also be transported into the aquatic environment through disposal of unused medicines, veterinary waste, or animal carcasses (Daughton and Ternes 1999; Boxall et al. 2004). The potential for a veterinary medicine to be released to the aquatic environment will be determined by several different criteria, including the method of treatment, agriculture or aquaculture practices, environmental conditions, and the properties of the veterinary medicine.

During the environmental risk assessment process for veterinary medicines, it is generally necessary to assess the potential for aquatic exposure to the product being assessed. For example, in the VICH phase I process, it is necessary to estimate aquatic exposure concentrations for aquaculture products, and during the phase II process it is also necessary to determine exposure concentrations for products used in livestock treatments. Assessment of exposure must take into account the many different pathways and scenarios that influence the transport of veterinary medicines into the aquatic environment. In some cases, we have a good understanding of how these exposure scenarios can be evaluated, whereas in other cases, there is insufficient knowledge to guide the exposure assessments. Therefore, in this chapter we evaluate the current state of our knowledge concerning exposure of veterinary medicines in aquatic systems and synthesize the available data on fate and transport. We have also identified gaps and uncertainties in our understanding of exposure in order to inform the regulatory community and identify research needs.

4.2 SOURCES OF VETERINARY MEDICINES IN THE AQUATIC ENVIRONMENT

From Chapter 2, it is clear that there are many potential sources of emission of veterinary medicines into the environment. This chapter focuses on direct or indirect pathways by which medicines can reach the aquatic environment. In the following sections, we review the inputs of veterinary medicines into our water resources, including both groundwater and surface water (Figure 4.1), through their use in agriculture and aquaculture.

4.2.1 TREATMENTS USED IN AGRICULTURE

The likelihood of exposures in the aquatic environment and the potential magnitude of these exposures will vary for different pathways (Table 4.1). However, the major route of entry into the environment is probably under conditions of intensive agriculture (Table 4.1, Section 1A). Veterinary medicines are excreted by the animal in urine and dung, and this manure material is collected and subsequently applied to agricultural land (Halling-Sørensen et al. 2001; and see Chapter 2).

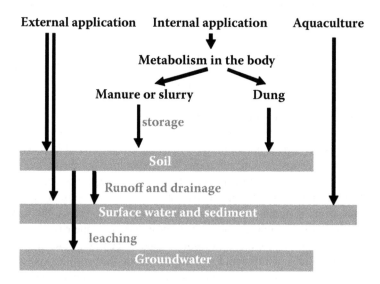

FIGURE 4.1   Direct and indirect pathways for the release of veterinary medicines into the aquatic environment.

Although each class of livestock production has different housing and manure production characteristics, the distribution routes for veterinary medicines are essentially similar. Following application onto soil, medicines may leach to shallow groundwater or be transported to surface water through runoff or tile flow (Hirsch et al. 1999; Meyer et al. 2000; Kay et al. 2004, 2005; Burkhard et al. 2005; Stoob et al. 2007). Potentially important releases into the aquatic environment can also occur when manure storage facilities overflow because of rain events or are breached by floods or when manure is accidentally spilled during storage or transport (Table 4.1, 2A). When manure is stored in lagoons, veterinary medicines may leach from these structures into groundwater or surface water (Table 4.1, 3A). The potential for impacts from manure spills or releases from lagoon sites should not be underestimated. For instance, in the state of Iowa in the United States, more than 1000 aerobic and anaerobic lagoons for manure storage and associated retention basins have been identified. The Department of Natural Resources in Iowa recorded 414 fish kills in the 10-year period between 1995 and 2002. These fish kills were thought to be related to spills during manure transport. These sources of veterinary medicines into the environment are not likely to be an important factor in product approvals, but they may be important considerations for product labeling or for the development of best management practices for manure storage and transport. Another significant but probably lower magnitude source of veterinary medicines is the deposition of urine and dung onto pasture land by animals that are being raised under low-density conditions (Table 4.1, 1B). Direct excretion of veterinary medicines in dung or urine into surface water may also occur when pasture animals have access to rivers, streams, or ponds (Table 4.1, 4B).

Inputs of substances that are applied and act externally may also be important (e.g., ectoparasiticides). Various substances are used externally on pasture animals, poultry, and companion animals for the treatment of external or internal parasites and infection. Sheep in particular require treatments for scab, blowfly, ticks, and lice that include plunge dipping, pour-on formulations, and the use of showers. The sheep dip products include insecticides from the pyrethroid (i.e., cypermethrin) and organophosphate (i.e., diazinon) classes. With externally applied veterinary medicines, both direct and indirect releases to the aquatic environment can occur (Table 4.1, 4B). Wash off of chemicals from the surface of recently treated animals to soil, water, and hard surfaces (e.g., concrete) may occur on the farm, during transport, or at stock markets (Littlejohn and Melvin 1991). Wash off of chemicals may also be a source of veterinary medicines from companion animals, although the magnitude of these releases is probably small (Table 4.1, 5C). In dipping practice, chemicals may enter watercourses following disposal of used dip and leakage of used dip from dipping installations (Table 4.1, 6A and 6B). Other topically applied veterinary medicines that are likely to wash off following use include udder disinfectants (containing anti-infective agents) for dairy cattle and endoparasiticides for treating cattle.

Contaminated water that was used to wash indoor animal holding facilities may be transported out of the farmyard or may be collected for later application to

TABLE 4.1

Major sources of veterinary medicines and the activities leading to exposure in aquatic environments

Activity	Source (animal — likelihood and magnitude)			VICH guidance scenario	Need for further guidance
	A: intensive	**B: pasture**	**C: companion animals**		
1) Direct excretion of manure from animal onto land, or land application of manure, litter, or compost (slurry and/or sludge) after collection or storage	C, Ho, P H5	C, P, Ho, S, E H3	X H1	Y (for intensive and pasture)	N
2) Manure spills, overflows during transport	C, Ho, P M/5	—	—	N	Y
3) Lagoon leakage, including runoff and transport to groundwater	C, Ho H2	—	—	N	Y
4) Direct excretion of dung and urine from animal into surface water	—	C, P, Ho, S, E M2	—	Y	—
5) Wash off of animals from external treatments (e.g. dips and pour-ons)	—	C,S L3	X L1	Y	—
6) Direct spillage of product and feeds containing product	C, Ho, P L2	C, P, Ho, S, E L1	—	N	N
7) Farm wastewater, wash waters, etc., that do not go to a lagoon	C, Ho, P, E M3	—	—	N	N
8) Runoff from hard surfaces: feedlots	C, Ho, P H5	—	—	Y	—
9) Runoff from hard surfaces: barnyards	C, Ho M4	C, S, E L2	X L1	Y	—
10) Wastewater treatment plants	S, C L1	—	X L1	N	N
11) Processing plant wastes	C, Ho, P, E H1	—	—	N	N

TABLE 4.1 (continued)
Major sources of veterinary medicines and the activities leading to exposure in aquatic environments

	Source (animal — likelihood and magnitude)				
Activity	A: intensive	B: pasture	C: companion animals	VICH guidance scenario	Need for further guidance
12) Disposal of inserts, containers in landfill, etc.	C, P, Ho, S, E L2	C, P, Ho, S, E L2	X L2	N	N

Note: Animal: C = cattle, Ho = hogs, P = poultry, S = sheep/goats, E = horses, X = companion animals, All = All animals. Likelihood of exposure: H = high, M = moderate, L = low. Magnitude of exposure: 5 (high) to 1 (low). The availability of exposure guidance (Committee for Medicinal Products for Veterinary Use [CVMP] 2006) is identified.

land (Table 4.1, 7A). In North America, intensive cattle production practices usually include housing of animals in feedlots for final weight gain prior to slaughter. The runoff of medicines from the hard surfaces of feedlots as a result of rain events may be a significant source of contamination of surface water (Table 4.1, 8A). Medicines washed off, excreted, or spilled onto farmyard hard surfaces may be washed off to surface waters during periods of rainfall (Table 4.1, 9A and 9B).

Other potential sources of contamination are emissions of dipping chemicals from wool-washing plants (Armstrong and Philips 1998) or emissions of therapeutic medicines from milk-processing plants. Wastewaters from these facilities are generally treated, but removal during treatment may not be adequate (Table 4.1, 10A). Veterinary medicines in the feces of companion animals that are deposited into domestic sewage may also be discharged from municipal treatment plants (Table 4.1, 10C). Although withdrawal periods are supposed to be sufficient to clear veterinary medicines from animal tissues, it is possible that liquid wastes from meat-processing plants may also contain these contaminants if wastewater treatment is not effective at removing these compounds (Table 4.1, 11A). Finally, the inappropriate disposal of containers and administration equipment (i.e., syringes and inserts) for veterinary medicines, or the deposition of these materials into landfills, could be a source to the aquatic environment (Table 4.1, 12A, 12B, and 12C).

4.2.2 TREATMENTS USED IN AQUACULTURE

The primary pathway for direct inputs of veterinary medicines to the aquatic environment is through intensive aquaculture. Like other forms of intensive food production, aquaculture will have environmental impacts, including high inputs of nutrients. Cultured fish and commercially important invertebrates (e.g., crustaceans and mollusks) raised in the crowded and stressful conditions of aquaculture are susceptible to epidemics of infectious bacterial, viral, and parasitic diseases. For example, salmon are prone to infection from parasitic sea lice that can have serious impacts on the health and marketability of the fish. Control of sea lice infestations requires good fish husbandry but frequently requires treatments with chemicals that are applied either by bath (immersion) or in medicated feeds. Antibiotics are used in both marine and freshwater aquaculture applications, with medicated feed being the primary mode of administration. However, fish can also be treated with antibiotics by immersion using soluble formulations. Infections of the integument and gills in freshwater fish are typically treated using baths with chemicals that are not specific to a target pathogen (e.g., hydrogen peroxide, potassium permanganate, or copper sulphate). Chemotherapeutic agents in baths may be released directly into the aquatic environment once the treatment is complete. A significant portion of the chemotherapeutics in medicated feeds may leave aquaculture facilities in feces or in surplus food (Lunestad 1992; Samuelsen et al. 1992a, 1992b). For example, certain antibiotics such as oxytetracycline are poorly absorbed by fish and are excreted largely unchanged in the feces. Thus, veterinary medicines may be present in water and sediment via surplus medicated feed or excretion by treated animals.

4.3 EXPERIMENTAL STUDIES INTO THE ENTRY, FATE, AND TRANSPORT OF VETERINARY MEDICINES IN AQUATIC SYSTEMS

4.3.1 AQUATIC EXPOSURE TO VETERINARY MEDICINES USED TO TREAT LIVESTOCK

Livestock medicines will either be excreted directly to soil or applied to soil in manure or slurry (see Chapter 2). Contaminants applied to soil can be transported to aquatic systems via surface runoff, subsurface flow, and drainflow. The extent of transport via any of these processes is determined by a range of factors, including the solubility, sorption behavior, and persistence of the contaminant; the physical structure, pH, organic carbon content, and cation exchange capacity of the soil matrix; and climatic conditions such as temperature and rainfall volume and intensity (Boxall et al. 2006). Most work to date on contaminant transport from agricultural fields has focused on pesticides, nutrients, and bacteria, but recently a number of studies have explored the fate and transport of veterinary medicines. Lysimeter, field plot, and full-scale field studies have investigated the transport of veterinary medicines from the soil surface to field drains, ditches, streams, rivers, and groundwater (e.g., Aga et al. 2003; Kay et al. 2004, 2005; Burkhard et al. 2005; Hamscher et al. 2005; Lissemore et al. 2006; Stoob et al. 2007). A range of experimental designs and sampling methodologies has been used. These investigations are described in more detail below and are summarized in Table 4.3.

4.3.1.1   Leaching to Groundwater

The movement of sulfonamide and tetracycline antibiotics in soil profiles was investigated at the field scale using suction probes (Hamscher et al. 2000a; Blackwell et al. 2005, 2007). In these studies, sulfonamides were detected in soil pore water at depths of both 0.8 and 1.4 m, but tetracyclines were not, most likely due to their high potential for sorption to soil. Carlson and Mabury (2006) reported that chlortetracycline applied to agricultural soil in manure was detected at soil depths of 25 and 35 cm, but monensin remained in the upper soil layers. There are only a few reports of veterinary medicines in groundwater (Hirsch et al. 1999; Hamscher et al. 2000a; Krapac et al. 2005). In an extensive monitoring study conducted in Germany (Hirsch et al. 1999), antibiotics were detected in groundwater at only 4 sites. Although contamination at 2 of the sites was attributed to irrigation of agricultural land with domestic sewage and hence measurements were probably due to the use of sulfamethazine in human medicine, the authors concluded that contamination of groundwater by the veterinary antibiotic sulfamethazine at 2 of the sites was due to applications of manure (Hirsch et al. 1999).

4.3.1.2   Movement to Surface Water

Transport of veterinary medicines via runoff (i.e., overland flow) has been observed for tetracycline antibiotics (i.e., oxytetracycline) and sulfonamide antibiotics (i.e., sulfadiazine, sulfamethazine, sulfathiazole, and sulfachloropyridazine), as reported by Kay et al. (2005), Kreuzig et al. (2005), and Gupta et al. (2003). The transport of these substances is influenced by the sorption behavior of the compounds, the presence of manure in the soil matrix, and the nature of the land to which the manure is applied. Runoff of highly sorptive substances, such as tetracyclines, was observed to be significantly lower than that of the more mobile sulfonamides (Kay et al. 2005). However, even for the relatively water-soluble sulfonamides, total mass losses to surface water have been reported to lie only between 0.04% and 0.6% of the mass applied under actual field conditions (Stoob et al. 2007). The presence of manure slurry incorporated into a soil matrix was observed to increase the transport of sulfonamides via runoff by 10 to 40 times in comparison to runoff, following direct application of these medicines to grassland soils (Burkhard et al. 2005). Possible explanations for this observation include physical "sealing" of the soil surface by the slurry or a change in pH as a result of manure addition that altered the speciation and fate of the medicines (Burkhard et al. 2005). It has been shown that overland transport from ploughed soils is significantly lower than runoff from grasslands (Kreuzig et al. 2005).

The transport of a range of antibacterial substances (i.e., tetracyclines, macrolides, sulfonamides, and trimethoprim) has been investigated using lysimeter and field-based studies in tile-drained clay soils (Gupta et al. 2003; Kay et al. 2005, 2004; Boxall et al. 2006). Following application of pig slurry spiked with antibiotics to an untilled field, test compounds were detected in drainflow at concentrations up to a maximum of 613 µg L^{-1} for oxytetracyline and 36 µg L^{-1} for sulfachloropyridazine (Kay et al. 2004). Spiking concentrations for the test compounds were all similar, so differences in maximum concentrations were likely due to differences in sorption behavior. In a subsequent investigation at the same site (Kay et al. 2004), in which the soil was tilled, much lower concentrations were observed in the drainflow (i.e., 6.1 µg L^{-1} for sulfachloropyridazine and 0.8 µg L^{-1} for oxytetracyline). Although the pig slurry used in these studies was obtained from a pig farm where tylosin was used as a prophylactic treatment, this substance was not detected in any drainflow samples, possibly because it is not persistent in slurry (Loke et al. 2000).

Once a veterinary medicine is introduced into the environment on a farm or in an aquaculture facility, there are many processes that will affect its fate in the aquatic environment, including partitioning, biological degradation, photolysis, and hydrolysis. These fate processes were reviewed by Boxall et al. (2004). Partitioning to organic material may limit bioavailability and influence environmental fate. The chemicals may enter aquatic systems in association with organic matter (dissolved or particulate) or in the aqueous (dissolved) phase. Many of the tetracycline antibiotics interact strongly with organic matter, which may limit their biological availability. The quinolones, tetracyclines, ivermectin, and furazolidone are all rapidly photodegraded, with half-lives ranging from < 1 hour to 22 days, whereas trimethoprim, ormethoprim, and the sulfonamides are not readily photodegradable (Boxall et al. 2004). Ceftiofur is one of the few veterinary compounds identified that is subject to rapid hydrolysis, with a half-life of 8 days at pH. Although propetamphos was rapidly hydrolyzed at pH 3, at environmentally relevant pH levels (6 and 9), hydrolysis of this compound was much slower.

Monitoring of streams and rivers in close proximity to treated fields has been performed to assess the potential for transport to receiving waters due to the inputs described above. In a small stream receiving drainflow inputs from fields where trimethoprim, sulfadiazine, oxytetracycline, and lincomycin had been applied, maximum concentrations ranged from 0.02 to 21.1 µg L^{-1} for sulfadiazine and lincomycin, respectively (Boxall et al. 2006). At this site medicines were also detected in sediment at concentrations ranging from 0.5 µg kg^{-1} for trimethoprim to 813 µg kg^{-1} for oxytetracycline. At a site where there was transport of veterinary medicines from agricultural fields by both drainflow and runoff, maximum concentrations of sulfonamides in a small ditch adjacent to fields treated with pig slurry ranged from 0.5 µg L^{-1} for sulfamethazine to 5 µg L^{-1} for sulfamethoxazole (Stoob et al. 2007). In a region of the Grand River system in Ontario, Canada, that passes through agricultural areas, Lissemore et al. (2006) detected several veterinary medicines at ng L^{-1} concentrations, including lincomycin, monensin, and sulfamethazine. The maximum mean concentration of monensin observed at a site in the Grand River was 332 ng L^{-1} (Lissemore et al. 2006).

4.3.1.3   Predicting Exposure

Guidelines are available on how to assess exposure to livestock medicines in aquatic systems (International Cooperation on Harmonization of the Technical Requirements for Registration of Veterinary Medicinal [VICH] 2004; Committee for Medicinal Products for Veterinary Use [CVMP] 2006) through the most common pathways. A number of approaches have been developed for predicting concentrations of veterinary medicines in soil, groundwater, and surface waters (e.g., Spaepen et al. 1997; Montforts 1999). Generally, at early stages in the risk assessment process, simple algorithms are used that provide a conservative estimation of exposure in soils. If an environmental risk is shown at this stage, more sophisticated models are used. An outline of a number of the different algorithms is provided below, and, where possible, we have tried to evaluate these against experimental data.

In order to estimate the concentrations of veterinary medicines in aquatic systems, a prediction of the likely concentration in soils is required as a starting point. Estimates of exposure concentrations in soil are typically derived using models and model scenarios. The available modeling approaches for estimating concentrations in soils are described in detail in Chapter 6 (Section 6.7).

Concentrations in groundwater ($PEC_{groundwater}$) and surface water ($PEC_{surface\ water}$) are estimated from the soil concentrations. Maximum concentrations in groundwater can initially be approximated by pore water concentrations (i.e., $PEC_{groundwater} = PEC_{pore\ water}$), which can be derived according to equations laid out in the guidelines for evaluating exposures to new and existing substances (CVMP 2006). Based on these pore water concentrations, surface water concentrations are approximated by assuming runoff and drainflow concentrations to equal pore water concentrations, and subsequently applying a dilution factor of 10 to simulate the dilution of these concentrations in a small surface water body (i.e., $PEC_{surface\ water} = PEC_{pore\ water}/10$). If these highly conservative approximations indicate a risk to the environment, more advanced models are recommended for calculating PECs in groundwater and surface water. Two modeling approaches have been recommended for use with veterinary medicines, namely, VetCalc and FOCUS (CVMP 2006). These are described in more detail below.

VetCalc (Veterinary Medicines Directorate n.d.) estimates PEC values for groundwater and surface water using 12 predefined scenarios in Europe, which were chosen on the basis of the size, diversity, and importance of livestock production; the range of agricultural practices covered by the scenarios; and distribution over 3 different European climate zones (Mediterranean, Central Europe, and Continental Scandinavian). Each of the scenarios has been ranked in terms of its potential for predicting inputs from specific livestock animals (e.g., cattle, sheep, pigs, and poultry). The model also includes the typical manure management practices for the region on which the scenario is based. The VetCalc tool addresses a wide variety of agricultural and environmental situations, including characteristics of the major livestock animals, associated manure characteristics, local agricultural practices, characteristics of the receiving environment (e.g., soil or water), and the fate and behavior of chemicals within 3 critical compartments (i.e., soil, surface water, and groundwater).

Background information on these key drivers is taken into account in each scenario within the model database. Based on the dosage regime and chemical characteristics, VetCalc first calculates initial predicted concentrations in soil and manure. These are then used to simulate transport to surface water through runoff and leaching to groundwater. A third, fugacity-based model simulates the subsequent fate in surface water.

Another suite of mechanistic environmental models and accompanying scenarios has been created by a working group in Europe known as the Forum for the Coordination of Pesticide Fate Models and Their Use (FOCUS n.d.) to simulate the fate and transport of pesticides in the environment. Groundwater calculations developed by FOCUS involve the simulation of the leaching behavior of pesticides using a set of 3 models (PEARL, PELMO, and MACRO) in a series of up to 9 geographic settings that have various combinations of crops, soils, and climate. Groundwater concentrations are estimated by determining the annual average concentrations in shallow groundwater (1 meter soil depth) for a period of 20 consecutive years, then rank ordering the annual average values and selecting the 80th percentile value for comparison with the $0.1\ \mu g\ L^{-1}$ drinking water standard that has been established by the European Union.

The surface water and sediment calculations are performed using an overall calculation shell called SWASH (surface water scenarios help) that controls 4 models that simulate runoff and erosion (pesticide root zone model, or PRZM), leaching to field drains (MACRO), spray drift (internal to SWASH), and, finally, aquatic fate in ditches, ponds, and streams (toxic substances in surface waters, or TOXSWA). These simulations provide detailed assessments of potential aquatic concentrations in a range of water bodies located in up to 10 geographical and climatic settings. FOCUS models were originally designed for exposure assessments of pesticides. However, the CVMP guidance document (2006) provides some recommendations on how the model can be manipulated for applications to veterinary medicines, although much more model validation is needed to assess model performance for veterinary medicines.

4.3.1.4   Comparison of Modeled Concentrations with Measured Concentrations

The relatively simple algorithms suggested by CVMP (2006) for predictions of PECs in groundwater and in surface water would be expected to yield conservative estimates of levels in the environment. To test this assumption, we compared measured environmental concentrations (MECs) for soil, leachate, runoff, drainflow, and groundwater from the semifield and field studies to PECs for soil, pore water, and surface water predicted according to the algorithms reviewed above.

Wherever possible, actual measured or spiked manure concentrations were used as the starting point for the calculation of soil concentrations. Also, where

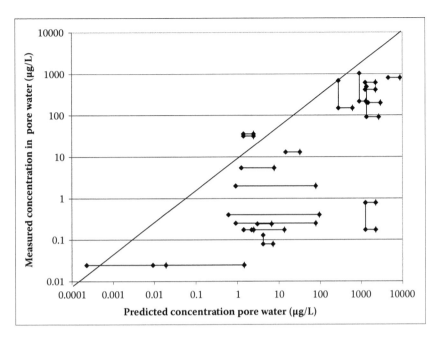

FIGURE 4.2   Comparison of predicted pore water concentrations with measured maximum concentrations in leachate, groundwater, drainflow, and runoff water for 8 veterinary medicines for which measured concentrations were available in field and semifield studies.

possible, actual depths of incorporation were used instead of the default value of 5 cm. In all other cases, default concentrations in manure for a given animal type and veterinary medicine had to be predicted from a knowledge of the treatment dosage and regime (Spaepen et al. 1997). Measured concentrations were either close to or significantly lower than the predicted concentrations, indicating that the models are indeed conservative (Figure 4.2). In those cases where manure loadings had to be estimated, the predicted soil concentrations were highly conservative. In those cases, where manure concentrations were either measured or spiked, there was better agreement between predicted and measured soil concentrations.

To see whether algorithms for aquatic PECs were also conservative, PECs in soil pore water were estimated using minimum and median K_{oc} values and then compared to measured concentrations in leachate, groundwater, drainflow, and runoff from 8 of the studies listed in Table 4.2. Again, the results show that the pore water PECs are usually conservative estimates of the measured concentrations (Figure 4.2). However, when measured concentrations in receiving waters are compared to surface water predictions derived from the pore water predictions, there were 3 instances where measured concentrations exceeded predicted concentrations (Figure 4.3). In all 3 cases, the substance belonged to the tetracycline group. This is in agreement with the findings of Kay et al. (2004) that indicate that strongly sorbing compounds such as tetracyclines can be transported bound to colloidal organic matter. This mode of transport is currently not

TABLE 4.2

Field scale and column studies reported in the literature on the fate and transport of veterinary medicines

Study	Location	Study substances	Study scale	Application	Manure type	Manure storage	Matrices analyzed	Sampling regime	Application rate	Soil data	Climate data
1) Aga et al. (2003)	Illinois, US	Tetracycline	Column	Natural	Pig	U	S, L	Set times	Y	Detailed	Continuous irrigation
2) Boxall et al. (2005)	Derbyshire, UK	Lincomycin Oxytetracycline Sulfadiazine Trimethoprim	Field	Natural	Pig	Y	S, SW	S: set times; SW: continuous	Calculated	Detailed	Y
3) Blackwell et al. (2007)	Derbyshire, UK	Oxytetracycline Sulfachloropyridazine Tylosin	Plot	Spiked (except tylosin)	Pig	Y	S, IW	Set times	Y	Detailed	Y
4) Burkhard et al. (2005)	Zürich, Switzerland	Sulfadiazine Sulfadimidine Sulfathiazole	Plot	Spiked manure or aqueous solution	Pig	U	OF	Continuous	Y	Detailed	irrigation
5) Gupta et al. (2003)	Minnesota, US	Tetracycline Chlortetracycline Tylosin	Plot	Natural	Pig	Y	OF, DW	Continuous	Y	Some data	N
6) Halling-Sørensen et al. (2005)	Askov and Lundgaard, Denmark	Chlortetracycline Tylosin	Field	Natural	Pig	Y	S	Set times	Calculated	Detailed	Y
7) Hamscher et al. (2005)	Lower Saxony, Germany	Tetracycline Chlortetracycline Sulfamethazine Sulfadiazine	Field	Natural	Pig	Y	S, GW	Set times	Y	Y	N

Study	Location	Compounds									
8) Mackie et al. (2006)	Illinois, US	Tetracycline Chlortetracycline Oxytetracycline Anhydrotetracycline B-apooxytetracycline Anhydrochlortetracycline	Field	Natural	Pig	Y			N	Limited	N
9) Kay et al. (2004)	Cestershire, UK	Oxytetracycline Sulfachloropyridazine Tylosin	Field	Spiked manure (except tylosin)	Pig	Y	DW, S	S: set times DW: continuous	Y (except tylosin)	Detailed	Y
10) Kay et al. (2005)	Cestershire, UK	Oxytetracycline Sulfachloropyridazine Tylosin	Lysimeter	Spiked (except tylosin)	Pig	Y	S.L	Continuous	Y (except tylosin)	Detailed	Y
11) Kay et al. (2005)	Cestershire, UK	Oxytetracycline Sulfachloropyridazine Tylosin	Plot	Spiked (except tylosin)	Pig	Y	OF	Continuous	Y (except tylosin)	Detailed	Y
12) Kreuzig and Holtge (2005)	Lower Saxony, Germany	Sulfadiazine	Plot and lysimeter	Spiked	Pig	NA	S, L	Set times	Y	Detailed	Irrigation
13) Kreuzig et al. (2005)	Lower Saxony, Germany	Sulfadiazine	Plot	Spiked	Pig	NA	OF	Continuous	Y	Detailed	Irrigation
14) Stoob et al. (2007)	Switzerland	Sulfamethoxazole Sulfadimethoxine Sulfamethazine	Field	Spiked manure (except sulfamethazine)	Pig	Y	SW	Continuous	Y	Detailed	Y

Note: Y = yes, N = no, M = manure, S= soil, IW = interstitial water, GW = groundwater, SW = surface water, DW= drainage water, OF = overland flow water, L = leachate, Se = sediment.

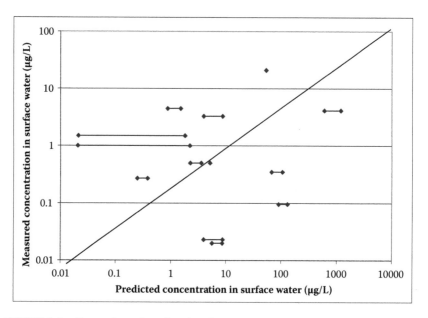

FIGURE 4.3 Comparison of predicted surface water concentrations with measured concentrations for surface water for 9 veterinary medicines for which measured concentrations were available in field studies.

considered in the simple algorithms suggested by CVMP (2006). Thus, in the case of strongly sorbing compounds, the algorithms may not provide a conservative estimate of the PEC.

VetCalc was also evaluated against measured concentrations. The persistence and K_{oc} values used in this evaluation are summarized in Table 4.3. VetCalc estimates of concentrations in soil were generally higher than measured soil concentrations under field application conditions (Figure 4.4). The only exception was tylosin, where the predicted soil concentration was 10 orders of magnitude lower than the measured soil concentration, which was 0.03 mg kg. The model assessment for tylosin considered degradation during storage and assumed a typical manure storage scenario, but it is possible that the field storage duration was significantly lower than the default value, explaining the higher measured concentrations.

For concentrations in surface water, with the exception of oxytetracycline, there was always at least 1 VetCalc scenario that predicted higher concentrations than the measured maximum concentrations (Figure 4.5). There were also always some VetCalc scenarios that resulted in predicted concentrations lower than measured concentrations. This is not perhaps surprising, as field studies are generally performed at sites that are known to be vulnerable to transport of chemicals to water, whereas VetCalc models the fate of substances across a range of European agricultural, soil, and climatic scenarios. For our case study compounds, the scenarios for Belgium, Denmark, Finland, France, Germany, and the United Kingdom tended to give estimates of surface water concentrations that were lower than

TABLE 4.3

Input data on chemical and physical parameters of veterinary medicines used in modeling exercises

Substance	CAS	Treatment group	Dose (mg kg⁻¹ d⁻¹)	Treatment duration (d)	Kd (L kg⁻¹)	Koc (L kg⁻¹)	DT50 (d)
Chlortetracycline	64-72-2	Hogs	20	7		4681-34270000 Median 400522	—
Enrofloxacin		Poultry	10	10	3037	186342	359-696
					5612	768740	
					1230	99975	
					260	16506	
					496	70914	
					6310	Median 99975	
					3548		
					4670		
					5986		
Lincomycin	154-21-2	Hogs	22	21		111	5.2
Monensin					9.3		7.4
							7.5
Oxytetracycline	6153-64-6	Hogs	20	15	680	42506	18
					670	47881	16
					1026	93317	
					417	27792	
						Median 47932	

(continued on next page)

TABLE 4.3 (continued)

Input data on chemical and physical parameters of veterinary medicines used in modeling exercises

Substance	CAS	Treatment group	Dose (mg kg⁻¹ d⁻¹)	Treatment duration (d)	Kd (L kg⁻¹)	Koc (L kg⁻¹)	DT50 (d)
Sulfachloropyridazine	80-32-0	Hogs	20	10	3.3	16	2.8
					8.1	18	3.5
						Median 17	
Sulfadiazine	68-35-9	Hogs	25	3		61	10.4
Sulfamethazine	57-68-1	Hogs				Min 46	
						Median 110	
Sulfathiazole	72-14-0	Hogs				116	
						176	
						80	
Sulfadimethoxine						Median 118	
						Min 89	
						Median 144	
Sulfamethoxazole						Min	
Tetracycline	60-54-8	Hogs	60	5	—	2723-65090000	
						Median 420999	
Trimethoprim	738-70-5	Hogs	8	5		1680-3990	110
						Median 2589	
Tylosin	1401-69-0	Hogs	25	3		200-7988	<2 (pig slurry)
						Median 1264	95
							97

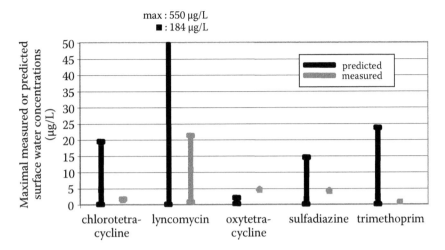

FIGURE 4.4 Comparison of VetCalc predictions of environmental concentration in soil (PEC$_{soil}$) under 12 scenarios with data on measured soil concentrations (MEC$_{soil}$).

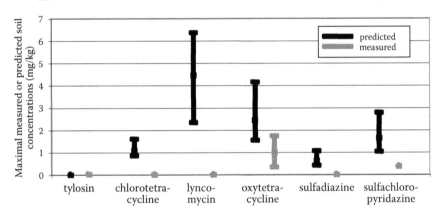

FIGURE 4.5 Comparison of VetCalc predictions of environmental concentration in surface water (PEC$_{surface\ water}$) under 12 scenarios with data on measured surface water concentrations (MEC$_{surface\ water}$).

the measured concentrations reported in the few studies on veterinary medicines in European surface waters. As with the simple algorithms, surface water concentrations of oxytetracycline were underpredicted, probably because colloidal or particle-bound transport is not currently considered by VetCalc.

4.3.2 AQUACULTURE TREATMENTS

Veterinary medicines are widely used in aquaculture. For example, it is estimated that more than 200 000 kg of antibiotics are used annually in US aquaculture (Benbrook 2002), with about 75% of the antibiotics administered in aquaculture entering the environment via excretion of feces and uneaten medicated feed

(Lalumera et al. 2004). The inputs and use vary between marine and freshwater facilities. It has been recently recognized that the prophylactic use of antibiotics in aquaculture is a growing environmental problem (Cabello 2006), particularly in developing countries.

Four general types of systems are used in aquaculture: ponds, net pen cage, flow-through systems (e.g., Figure 4.6), and recirculating systems. The potential exposure pathways differ between the systems. Floating and bottom-culture systems are also used for culturing of mussels, clams, and oysters, but medicines are rarely used to treat these organisms. In each of these systems there are 2 major sources of medicine release: emissions from bath treatments or medicated feeds.

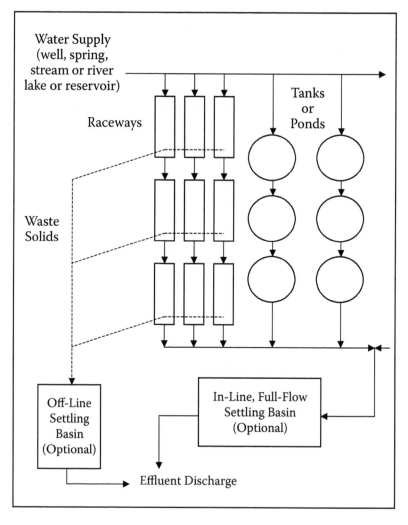

FIGURE 4.6 Schematic of a typical flow-through aquaculture facility showing the basic and optional components of the system.

Baths can be either static or flow-through, depending on the type of aquaculture system and species being raised. Detailed information on the construction, operation, and maintenance of these different aquaculture facility types can be found elsewhere (e.g., Lazur and Britt 1997; Losordo et al. 1999; Mazik and Parker 2001; Tucker et al. 2001; Chen et al. 2002; Hargreaves et al. 2002; Steeby and Avery 2002; Whitis 2002; Stickney 2002; US Environmental Protection Agency 2004).

4.3.2.1   Inputs and Fate of Marine Aquaculture Treatments

Both antibiotics and sea lice treatments are used in marine aquaculture. Sea lice treatments include the organophosphates (azamethiphos), pyrethroids (cypermethrin and deltamethrin), hydrogen peroxide, avermectin compounds (emamectin benzoate), and chitin synthesis inhibitors (teflubenzuron and diflubenzuron). Depending on the class, these may be administered either as a bath treatment or as additives in medicated feed. Bath treatments are conducted by reducing the depth of the net in the salmon cage, thus reducing the volume of water. The net pen and enclosed salmon are surrounded by an impervious barrier, and the chemical is added to the recommended treatment concentration. The salmon are maintained in the bath for a period of 30 to 60 minutes, and then the barrier is removed and the treatment chemical is allowed to disperse into the surrounding water. Medicated feeds are prepared by adding concentrated mix containing the active ingredient to the feed during commercial preparation. The therapeutic agent is absorbed from the feed into the fish and is then transferred to the sea lice as they feed on the skin of the salmon. Medicated feeds are the primary method used to control sea lice in salmon aquaculture because of ease of use, safer handling by aquaculture personnel, and lower potential for losses to the environment (Burka et al. 1997; Alderman and Hastings 1998; Haya et al. 2005).

Avermectins are often used in medicated feeds because of their efficacy and low cost. The avermectin compound that is licensed for use in sea lice control is emamectin benzoate. Avermectins can reach the marine environment in uneaten feed pellets, or in the feces or biliary products excreted by fish. Emamectin benzoate is relatively persistent, is hydrophobic, and has the potential to adsorb to particulate material and marine sediments (Scottish Environmental Protection Agency [SEPA] 1999; Haya et al. 2005). In a field trial conducted in Scotland (SEPA 1999), this compound was occasionally detected in water samples at concentrations of up to 1.06 µg L^{-1}, but it was detected frequently in sediment samples near the salmon cages at concentrations up to 2.73 µg kg. This compound and its metabolites were detected in sediments up to 12 month post treatment.

A small number of antibiotics are registered for use in the fish aquaculture industry in Canada, the United States, and northern Europe. These include amoxicillin, florfenicol, and substances from the quinolone, fluoroquinolone, sulfonamide (including potentiated sulfonamides), and tetracycline classes. Both amoxicillin and florfenicol degrade rapidly in the environment. In contrast, substances from the quinolone groups have been detected around aquaculture facilities. For example, in a study conducted off the southwest coast of Finland, residues of oxolinic acid were detected in anoxic sediments collected below net pens at concentrations up to 0.2 mg kg^{-1} at 5 days posttreatment (Björklund et al. 1991). Oxytetracycline has been widely studied in terms of its environmental fate and persistence. The absorption rate of oxytetracycline across the gut wall in salmon is low (< 2% of the administered dose), and therefore fecal matter would be expected to contain high concentrations of antibiotics (Samuelson et al. 1992a; Weston 1996). Unconsumed antibiotic-treated feed pellets will be deposited directly below the pen site or, in high current areas, may be distributed more broadly. Mass balance budgets for oxytetracycline in the vicinity of salmon farms have shown that 5% to 11% of the total oxytetracycline input could be accounted for in sediment residues (Björklund et al. 1990; Coyne et al. 1994; Capone et al. 1996). From these data, it appears that most of the excreted oxytetracycline partitions into the dissolved and particle-associated phases of the water column. However, no study has directly measured the distribution of oxytetracycline in water around an aquaculture site following feed application.

Accumulation of antibiotics in sediments can occur either by direct deposition of treated feed in the vicinity of net pens or by adsorption of antibiotics onto settling particles (Pouliquen et al. 1992). For example, concentrations of oxytetracycline measured in coastal marine sediment at pen sites varied from < 10 mg kg^{-1} (Björklund et al. 1991) to a maximum of 240 mg kg^{-1} (Coyne et al. 1994). This antibiotic has also been detected in anoxic sediments near net pens in Norway and Finland for periods of more than 1 year after treatment (Björklund et al. 1991). The half-life of oxytetracycline in sediment was prolonged to 419 days under stagnant, anoxic conditions (Björklund et al. 1990).

4.3.2.2 Freshwater Aquaculture

There is a variety of veterinary medicines used in freshwater aquaculture, although compared to marine aquaculture there has been little research examining the environmental occurrence of veterinary medicines following use in freshwater aquaculture. Most research has focused on determining concentrations in water discharged or adjacent to fish aquaculture operations that have used antibiotic treatments (Smith et al. 1994; Bebak-Williams et al. 2002; Dietze et al. 2005), with some examination of concentrations in sediment (Lalumera et al. 2004; Bebak-Williams et al. 2002) and tissues (Xu et al. 2006; Wrzesinski et al. 2006). For example, Dietze et al. (2005) reported that maximum antibiotic concentrations in water reached 36 µg L^{-1} during treatment and remained detectable for up to 48 days following treatment. These concentrations were similar to concentrations found in pig slurry lagoons (Meyer et al. 2003), so it is obvious that freshwater aquaculture has the potential to be an important source for the release of antibiotics into the aquatic environment. Preliminary results indicate that more frequent and higher antibiotic concentrations may be found in water from intensive aquaculture facilities, relative to less intensive hatcheries (Dietze et al. 2005). Antibiotics could accumulate in fish tissues, water, and sediment to a greater extent in recirculating systems (Bebak-Williams et al. 2002).

4.3.2.3   Modeling Exposure from Aquaculture Treatments

Although exposure assessment approaches are available for estimating environmental concentrations of aquaculture treatments (e.g., VICH 2004; CVMP 2006), these are not well developed. For example, it is currently recommended for phase I assessments (CVMP 2006) that the PEC be estimated by calculating the total amount of active ingredient that is added to an aquaculture system and then subtracting the amount that is retained in "sludge" (i.e., waste material that is filtered or settles out within the facility). This calculation is not appropriate for assessing aquatic exposures under many aquaculture scenarios, such as exposures in net pens. In addition, limited guidance is available for higher tier assessments of products intended for use in aquaculture.

Therefore, in this section, several simple algorithms are proposed for calculating "generic" initial predicted environmental concentrations ($PEC_{initial}$, also known as environmental introduction concentrations) for veterinary medicines applied in baths or in medicated feeds in the 4 general types of aquaculture systems described earlier. For closed or self-contained facilities, these PEC values represent the concentrations of the veterinary medicine expected in effluents at the point of release or discharge to surface water. For open systems, such as marine net pens, the PECs represent concentrations at points immediately adjacent to the treatment area that may disperse laterally and vertically to a wider environment. Guidance is also provided on ways to refine exposure assessments using medicine-specific and/or facility-specific data.

Because of the wide variability in the design and operation of different aquaculture facilities it is preferable, when possible, to develop a series of facility-specific PECs for use in risk assessment. Unfortunately, it is difficult to do this, particularly for preapproval assessments of new medicines, because of the large number of potential facilities, the lack of facility-specific data, and the need to approve medicines on a country- or region-wide basis. However, in some cases, survey data may be available for representative aquaculture facilities that would be expected to use a medicine, once approved, or for facilities that are using a particular medicine while it is undergoing investigational use, prior to approval. These data could include such things as flow rates, treatment intervals, tank and raceway sizes, solids or medicine removal rates, and surface water dilution factors. These data may be used to develop a range of PECs, and in some instances may allow for the development of probabilistic exposure assessments (see Chapter 3).

We do not recommend a specific default dilution factor, or factors, for calculating the $PEC_{sw-initial}$ (SW = surface water). Dilution factors representing the ratio of the combined flow rate (volume and time) of the receiving water and the effluent discharge, divided by the flow rate of the effluent discharge alone, may range from 1 (no dilution) for effluent-dominated headwaters to 1 million or more for large rivers. For an initial assessment, it is suggested that "reasonable worst-case" scenarios be developed to determine appropriate, but conservative, dilution factors for each of the aquaculture systems in which use of the medicine occurs or is expected to occur. The location of use and type of receiving water (stream, river, lake, estuary, etc.) will be the most important factors to consider in developing these scenarios. Dilution will often vary significantly with the season and weather, so consideration should be given as to when the discharge will most likely occur. This will depend on the medicine, what it is used for (e.g., species and indication), where it is used, and when it is most likely to be used. For example, unintentional discharges, such as those due to flooding of ponds, are most likely to occur during periods of high rainfall when flow rates in receiving waters will also be high. In contrast, medicines used in flow-through systems are more likely to be discharged year-round, including during periods of low flow.

We caution against using a default dilution factor for the calculation of $PEC_{sw\text{-}initial}$ without first consulting the appropriate regulatory authority for information on effluent discharges. Regulations in some jurisdictions, such as certain states in the United States, do not allow any toxicity in the mixing zone where an effluent discharges to and mixes with surface water. This means that dilution cannot be considered and assessments must be based on concentrations at the end of the pipe, where the effluent discharges.

4.3.2.3.1   Pond Systems

Medicated feeds and bath (immersion) treatments are both used to administer medicines to aquaculture species reared in closed ponds, which include most levee and watershed ponds that are operated as closed (static) systems with intermittent flow during filling and draining operations. Aquaculture ponds may also be operated as open systems with a continual inflow and outflow. Calculations for "open" ponds are addressed in the section on flow-through systems. Whole-pond bath treatments are not usually an economical alternative for most aquaculture medicines; therefore, most treatments are made via medicated feeds. Exceptions include oxidants such as potassium permanganate, metallic salts such as copper sulfate, and parasiticides such as formalin. Some of these compounds may be classified as medicines, pesticides, biocides, or disinfectants, depending on the jurisdiction and their intended use.

The release of veterinary medicines from aquaculture ponds is usually intermittent or irregular and may be either controlled (e.g., due to draw-down for harvesting or cleaning) or uncontrolled (e.g., through overtopping of dams or levees during flood conditions). The magnitude of the release will depend on several factors, including the type of medicine treatment (feed versus bath), the persistence of the medicine in pond water, and the time of the discharge in relation to the time of the treatments. In most cases, the time of discharge will be well after the time of treatments. However, because the discharge is not always controllable, it is recommended that the $PEC_{initial}$ be conservatively calculated under the assumption that the entire amount of medicine originally applied to the pond is present in the water column at the time of discharge.

4.3.2.3.2   Pond Systems with Bath Treatments

For levee ponds with bath treatments, the $PEC_{initial}$ is simply the treatment concentration (as active ingredient, or a.i.) in mg L^{-1} (ppm). In most cases, this concentration is specified on the medicine product label. If not, it may be calculated as follows:

$$PEC_{initial} = \frac{M \times 1,000,000 \times P}{100 \times V} \qquad (4.1)$$

where

M = mass of medicinal product added to pond (kg)
$1\,000\,000$ = conversion factor (kg to mg)
P = percentage of active ingredient in medicine (w/w)
100 = conversion factor (percentage to fraction a.i.)
V = volume of pond (L)

For information on methods for determining the volumes of ponds, consult SRAC Publication No. 103 (Masser and Jensen 1991).

For aquatic life in receiving waters, the $PEC_{sw-initial}$ is determined from the $PEC_{initial}$ by taking into account dilution in the receiving water, but assuming no other degradation or dissipation (e.g., adsorption) of the medicine prior to discharge from the pond.

$$PEC_{sw-initial} = \frac{PEC_{initial}}{Dilution\ Factor}$$

For watershed ponds undergoing bath treatments, the $PEC_{initial}$ is the treatment dose in mg L^{-1}, adjusted for the potential inflow and outflow of pond water prior to discharge. In most cases this is probably not significant, so the same algorithms used above for the levee pond scenario with a bath treatment may be used here to calculate the $PEC_{initial}$ and $PEC_{sw-initial}$.

4.3.2.3.3 Pond Systems with Feed Treatments

With a medicated feed treatment used in a levee pond, the concentration in water is estimated at the end of the treatment period, when it is expected to be highest. The conservative default assumption for an initial assessment is that 100% of the medicine that is initially present in medicated feed is subsequently released to the water column (within the treatment period) through a combination of leaching from feed and uptake and excretion by the animals being treated. It is also assumed that there is no other degradation or dissipation of the medicine prior to discharge from the pond. The PEC calculation is as follows:

$$PEC_{initial} = \frac{(D \times BW \times N \times f) - L}{V} \qquad (4.2)$$

where

D = dose of the active ingredient (mg kg^{-1} body weight day^{-1})
BW = body weight of all animals being treated (kg)
N = number of days of medicated feed treatment
f = fraction of medicine metabolized in fish
L = feed lost to sediment
V = volume of pond (L)

The daily treatment dose of the active ingredient (mg kg^{-1} body weight day^{-1}) is usually specified on product labeling, but can be calculated based on feeding rate information for the species and life stage being treated if the concentration of the medicine in the feed is known. Fish are typically fed a percentage of their body weight each day, which may vary from < 1% to 10% or more depending on the species, size of fish, water temperature, and other factors. For catfish, daily feed requirements range from 1.2% body weight per day (% BW) for a 500 g fish at 22.8 °C to 3.0% BW for a 20 g fish at 20.0 °C (Westers 2001). Publications by Westers (2001), Huet (1994), Shepherd and Bromage (1992), or other experts can be consulted for species-specific information.

The PEC is directly proportional to the biomass of animals in the pond, which is often expressed in terms of density (e.g., kg m^{-3}, kg ha^{-1}). Density will vary depending on the species, size, time of year, and other factors such as whether or not there is supplemental aeration. In general, pond systems cannot support nearly the same densities as flow-through systems because dissolved oxygen will become limiting as the density increases. Fish densities in closed ponds may range from about 0.05 kg m^{-3} up to 2 kg m^{-3}, depending on the amount of fertilization, supplemental feeding, and aeration (Westers 2001). Because small fish have a higher metabolic rate and consume more oxygen per unit of body weight than large fish, they cannot be raised to as high a density. For example, fingerling catfish are raised to a density of 0.33–0.67 kg m^{-3} in 1-meter-deep ponds, whereas adult catfish are raised to a density of 1.1 kg m^{-3}.

If adequate and reliable data are available, the PEC$_{initial}$ may be adjusted by taking into account the amount of feed consumed compared to the total amount fed. This should only be done if the medicine is not very soluble in water and is unlikely to leach from the uneaten feed back into the water column. Adsorption, metabolism, and excretion data for the medicine in the species being treated may also be used to adjust the PEC$_{initial}$ if these data are available, and as long as there is adequate information to indicate that the metabolites have significantly reduced toxicity compared to the parent compound. If data on the metabolites are not available, it is generally assumed that they are just as active as the parent, and a total residue approach is used to calculate the PEC$_{initial}$. Some veterinary medicines, including many antibiotics, are poorly absorbed in the gut and are largely excreted unchanged in the feces. In this case, it is generally assumed that the medicine will leach from the feces once excreted and will contribute to the PEC$_{initial}$.

For aquatic life in receiving waters, the PEC$_{sw-initial}$ is determined from the PEC$_{initial}$ by taking into account dilution in the receiving water, in the same way as described previously for levee ponds.

Algorithms for a watershed pond with feed treatments are the same as shown above for the levee pond with feed treatment. In theory, it may be possible to adjust the initial PEC values by taking into account the volume of water flowing into and out of the watershed pond during the period of treatment with medicated feed. However, in practice this is very difficult to do because the flow rate will depend on the amount of local runoff to the pond, which in turn will depend on the watershed-to-pond area, the amount of precipitation and evaporation, and watershed characteristics such as slope, type and extent of vegetative cover, soil type, and antecedent moisture.

4.3.2.3.4   Net Pen and Cage Systems

Most aquaculture operations using net pens and cage systems are located in coastal marine waters, or in large freshwater lakes and reservoirs using floating enclosures. Atlantic salmon are the most common species reared in these systems worldwide; however, significant production of other species also occurs in these systems on a local basis (e.g., yellowtail and red sea bream in Japan). In the future, greater use of these systems is expected in off-shore and deep-water environments as the technology advances. In order to minimize storm damage, most of these off-shore systems will be submerged, or anchored on the seabed.

4.3.2.3.5   Net Pen and Cage Systems with Bath Treatments

In open water systems, bath treatments of fish in individual net pens are made using an impermeable barrier or liner (e.g., tarpaulin) to hold the medicine during treatment. The liner is placed outside of the net pen, and then both it and the net pen are raised until the fish are confined to a small area. The amount of applied medicine is based on the volume of the confined area. Once the treatment period has ended, the net pen and liner are lowered back into the water, the liner is removed, and the solution of medicine is allowed to disperse by the action of tide, waves, and currents. Treatments are usually made 1 pen at a time and as needed. This type of treatment is most common for control of ectoparasites such as sea lice, but may be effective for external bacterial and fungal diseases.

The $PEC_{initial}$ for this scenario is based on the volume of a single net pen, which is considered to be the location from which the medicine is released to the greater environment. Therefore, the $PEC_{initial}$ is the medicine concentration in the treated volume (i.e., enclosed in the barrier) after dilution into the total volume of the net pen in the lowered position. The equation described above for a levee pond with a bath treatment may be used to calculate the $PEC_{initial}$, except in this case the volume of the net pen is substituted for the volume of the pond. Information on the amount (kg) of medicine applied in the confined area during treatment is needed in order to calculate the $PEC_{initial}$. This can be calculated knowing the treatment concentration and volume of the confined area. A water depth of 3 m for the confined area during treatment may be assumed if specific data are not available.

To determine the $PEC_{sw-initial}$, dilution of the medicine is taken into account assuming a water column mixing zone that includes the area within and extending laterally some distance beyond the perimeter of the net pen in all directions on the surface and vertically down to the sea floor and water column interface. According to the permits for Atlantic salmon aquaculture issued by the Department of Environmental Protection for the state of Maine, United States, the lateral distance beyond the net pen perimeter is stipulated to be 30 m. This distance is based on requirements that the discharges from salmon aquaculture facilities should not cause conditions that are toxic to aquatic life outside of the allocated mixing zone. In the absence of other site-specific information, this 30-m lateral distance is a reasonable default value for calculating the size of the mixing zone. Other jurisdictions may allow for other sizes of mixing zones, so this value should be adjusted as appropriate. For example, SEPA defines the allowable zone of effect for azamethiphos as the lower of either 0.5 km² or 2% of the loch area.

There is no "standard" size of net pen, as this varies considerably from place to place. Therefore, data are needed to determine the appropriate pen size for the species being reared and treated and the locality of medicine use. When calculating the $PEC_{initial}$, it should be assumed that the water depth below the bottom of the net pen is 10 m, unless site-specific data are available. If it is likely that there will be multiple net pen treatments at a single aquaculture facility over a short period of time (e.g., several hours), consideration should be given to evaluating the treatments as additive exposures, particularly if the treatments are on net pens in close proximity to each other. If this is done, the size and volume of the mixing zone may need to be adjusted accordingly.

4.3.2.3.6 Net Pen and Cage Systems with Feed Treatments

Exposure to feed treatments in net pens and cages can be predicted in a similar manner to the pond scenario except that there is no defined volume of dilution such as the pond itself. For calculation of the $PEC_{initial}$ it is conservatively assumed that the entire medicine dose is released into the water column and is diluted into the volume of the treated net pen(s) or cage(s).

For calculation of the $PEC_{sw-initial}$, further dilution is assumed into a volume extending 30 m beyond the perimeter of the net pen(s) or cage(s) in all directions on the surface and to a depth of 10 m, or in accordance with local jurisdictional allowances. This calculation does not account for the movement of water and further dilution of the medicine that could occur over the treatment period, which is typically several days or more. If data are available, it may be possible to determine the effects of currents and tides on dispersion of the medicine and the resulting PEC_{sw}. Mean current speeds of 0.1 and 0.05 m s⁻¹ have been used for sites with high and intermediate dispersion, respectively (SEPA 2003). Modeling approaches used by SEPA for medicated feeds are discussed in a later section in this chapter on determining the $PEC_{sw-refined}$. Further refinement of the PEC_{sw} may take into account the pharmacokinetics of medicine absorption and elimination, and the disposition of the medicine in uneaten feed and in feces.

Many medicines will remain largely in uneaten feed and feces and therefore will be initially deposited on the bottom of aquatic ecosystems. From there they may leach back into the water column or become incorporated into sediment. Effects on benthic organisms are initially assessed using a worst-case PEC for sediment. The $PEC_{sediment-initial}$ is calculated by assuming that the entire amount of medicine that is originally present in feed is subsequently incorporated into the sediments under the treated net pen(s) or cages(s). An incorporation depth of 5 cm and a sediment density of 2400 kg m⁻³ wet weight are assumed for this calculation:

$$PEC_{sediiment-initial} = \frac{D \times BW \times N}{(A \times I) \times D_{sed}}$$

(4.3)

where

D = dose of the active ingredient (mg kg^{-1} body weight day^{-1})

BW = body weight of animals being treated (kg)

N = number of days of medicated feed treatment

A = area of treated net pen(s) or cage(s) (m^2)

I = incorporation depth in sediment (0.05 m)

D_{sed} = density of sediment, wet weight basis (2400 kg m^{-3})

As discussed previously, the total amount of medicine used in the treatment will depend on the biomass being treated, which in turn is dependent on the volume of the net pens and the density of fish being reared. Atlantic salmon smaller than 2.0 kg in size are usually reared at densities of less than 15 kg m^{-3}, whereas those larger than this may be raised to a density of 30 to 40 kg m^{-3} (Willoughby 1999).

4.3.2.3.7 *Flow-Through Systems with Bath Treatments*

There are two basic scenarios for bath treatments in flow-through systems depending on whether it is possible to stop the flow in the culture unit. The first is a flow-through exposure in which the medicine is continuously added to the treatment unit (e.g., raceway and tank) during the entire period of treatment. Note that in some cases using this treatment method, a large bolus of the medicine is added to the unit at the start of the treatment period to quickly bring the concentration up to the target level. The second bath treatment scenario is a static exposure in which the medicine is added and mixed after flow through the unit has been temporarily stopped. Flow is resumed after the treatment period is over. For both scenarios the PEC values are based on the total amount of medicine (as active ingredient) that is added to the system during treatment. In most facilities, and particularly those with multiple culture units in parallel, the medicine will undergo extensive dilution prior to discharge and will be released over an extended period, even if the actual treatment period is short (e.g., < 1 hour). The period of discharge is extended further if the aquaculture facility uses an in-line settling basin or a pond for solids removal. Because of this, the PEC values are normally expressed as a time-weighted average (e.g., 24-hour average).

In some cases, it may be desirable to estimate a short-term "peak" PEC value for the medicine if it is assumed that it is discharged in a plug. Studies of medicine dissipation in raceways indicate that once treatment is ended, the majority of the medicine is flushed out during the time period required for 2 volume exchanges of the raceway (Gaikowski et al. 2004). If this time period is used as the short-term averaging period for the "peak" PEC, it may be calculated as follows:

$$\text{PEC}_{\text{peak}} = \frac{M_{a.i.}}{F \times T} \tag{4.4}$$

where

$M_{a.i.}$ = mass of medicine applied (mg as active ingredient)

F = average water flow rate for entire aquaculture facility (L min^{-1})

T = time for 2 volume exchanges of the treatment unit (min)

The time-averaged PEC$_{initial}$ over a 24-hour period is calculated using this equation

$$PEC_{initial\text{-}24\,hr\,Avg} = \frac{C \times V}{D + P} \tag{4.5}$$

where

C = medicine concentration in bath treatment (mg L^{-1} as active ingredient)

V = total facility daily treated volume (L)

D = total facility effluent discharge volume over 24 hours (L)

P = in-line settling pond or basin volume (L)

This equation can account for treatment of multiple culture units during a single day under either static or flow-through treatments. For static treatments, V is estimated by multiplying the number of culture units that are expected to be treated by the volume of these units. For flow-through treatments, V is determined by multiplying the number of treated culture units by the flow rate to the culture unit by the duration of the bath treatment. Parameter D may be based on the lowest typical daily flow rate for the facility or the typical flow rate during the expected time of treatment (e.g., spring or fall).

Time-averaged PECs for longer periods may be developed by modifying the parameters of the 24-hour equation above. For example, the 96-hour average PEC$_{initial}$ is calculated by changing parameter F so that it represents the total facility discharge over 96 hours, rather than the flow for 24 hours. If the medicine is to be administered multiple times during a treatment regimen (e.g., a total of 5 treatments, with each made every other day), parameter V should be modified to account for the additional treatments that would occur during the period of interest for PEC averaging. For example, if a medicine is administered by bath every other day for a total of 3 times, and the averaging period for the PEC is 96 hours, parameter V should be multiplied by a factor of 2 because there will be 2 treatments during this period.

The time-average equation given above requires facility-specific data, such as the number of daily treatments and the volume of treated culture units. If these data are not available, a generic worst-case exposure scenario may be used to calculate the PEC$_{initial}$. One way to do this is to assume that the facility has only a single culture unit. The facility flow rate is therefore the same as the flow rate for the culture unit. This flow rate can be estimated from the maximum density (kg m^{-3}) expected for the species being cultured by using an appropriate flow index or other information from the scientific literature. The flow index is an empirically derived, species-specific value that represents the maximum weight of fish of any size that can be reared in a raceway with a constant flow of water. The flow index is expressed as follows:

$$F = \frac{W}{L \times I} \tag{4.6}$$

where

F = flow index
W = weight of fish (kg)
L = length of fish (cm)
I = water flow rate (m^3 min^{-1})

Most US government hatchery facilities are operated with a flow index in the range of 10 to 25 when using metric units (Mazik and Parker 2001). Hatchery efficiency declines at flow indices of less than 10, and water quality deteriorates at indices greater than 25.

Raceways for salmonids are usually designed for fish loads of 32 to 48 kg m$^-$. If we assume an arbitrary raceway volume of 100 m^3 and a density at the upper limit (i.e., 48 kg m^{-3}), the raceway would hold 4800 kg of fish. Assuming a fish load of 1.0 kg L^{-1} for salmonids (Mazik and Parker 2001), the flow rate for the raceway would be 4800 L min$^-$. Using this flow rate and the raceway volume, the time-weighted PEC$_{initial}$ can be calculated using Equation 4.6.

The PEC$_{sw-initial}$ for this scenario is calculated in the same manner as previously described for pond treatments, taking into account dilution in the receiving water, if applicable, and if this is allowable within a specific jurisdiction.

4.3.2.3.8　Flow-Through Systems with Feed Treatments

Medicated feeds are usually fed in flow-through systems over a period of 5 to 28 days, and therefore the PEC for a flow-through scenario is usually expressed as a time-weighted average for the period of treatment. In this case, the time-weighted average PEC is determined as follows:

$$\text{PEC}_{\text{initial-average}} = \frac{D \times BW \times N}{V} \tag{4.7}$$

where

D = dose of the active ingredient (mg kg^{-1} body weight day^{-1})
BW = total body weight of animals being treated (kg)
N = number of days of medicated feed treatment
V = total facility effluent discharge volume over the treatment period (L)

The conservative default assumption is that 100% of the medicine that is initially present in medicated feed is subsequently released to the water column (within the treatment period) through a combination of leaching from feed and uptake and excretion by the animals being treated. It can also be assumed that there is no other degradation or dissipation of the medicine prior to discharge. Another default assumption is that all of the animals within the aquaculture facility are given the medicated feed. If this is not likely to be the case, the BW parameter needs to be modified to take into account the number of culture units that will be treated and the weight of fish within these units. Weights may be estimated from density information and unit volumes, as described for the flow-through bath treatment scenario above.

At times, the PEC averaging period may not be the treatment period, but a period relevant for risk assessment (e.g., 96 hours or 21 days) in order to match the length of acute or chronic toxicity tests conducted for the medicine in question. In this case, the value for parameter V may reflect the length of the relevant time period.

Adsorption, metabolism, and excretion data for the medicine in the species being treated may also be used to adjust the $PEC_{initial}$ if these data are available; however, a total residue approach is the default unless there is adequate documentation to conclude that metabolites are much less toxic than the active ingredient.

4.3.2.3.9   Recirculating Systems

Recirculating systems are essentially a subset of flow-through systems in which the majority of water is not discharged. It is not possible to construct equations easily to calculate the PECs for these systems because the discharge concentration will depend on both the percentage of recirculation in the facility and the percentage of medicine removal in the biofilter prior to discharge. As a worst-case scenario for bath treatments, it can be assumed that all flow to the treatment unit is temporarily diverted to waste during the treatment period in order to avoid damage to the biofilter. Under these conditions, the equations presented for the flow-through scenario can be used to estimate the $PEC_{initial}$.

4.3.2.3.10   Refining PECs for Aquaculture Medicines

The $PEC_{sw\text{-}initial}$ may be further refined based upon additional site-specific considerations, such as treatment regimes (number of culture units treated, timing of treatments, and frequency of treatment), flow conditions, raceway sizes, and the use of a settling pond or presence of a treatment system, at individual hatcheries. Refinements that include factors for dilution within the facility and dilution in the surrounding aquatic environment may also be considered. Additional refinement may be made to account for the physical, chemical, and biological properties of the medicine that affect environmental fate and transport, both within the aquaculture facility and within the receiving stream. Due to the complexity and site-specific nature of these factors, not every combination of possible refining factors will be presented. Important factors for each of the aquaculture systems are discussed below.

Whether treatments are by bath or feed, an important consideration in a pond system is the timing of the treatment, relative to the timing of release from the pond. If this interval is long, as is typical for levee ponds, and the medicine degrades, the $PEC_{sw\text{-}refined}$ may be reduced accordingly. For watershed ponds, a heavy rain occurring close to the time of use of the medicine will result in a worst-case distribution of the medicine downstream. The distribution of the medicine within the pond will depend upon the partitioning between the water, the suspended solids, and the sediment, which may be predicted through knowledge of the K_{oc}, and dissolved organic carbon (DOC) and suspended solids concentrations. When the pond is drained, some of the medicine will remain in the water column and some will be sequestered in the sediment. The fate of the medicine in the pond will also be influenced by various degradation processes, such as hydrolysis, photolysis, and biodegradation. Data from laboratory fate studies, such as OECD 308, which examines the degradation of chemicals in a water and sediment system, can be used to estimate these processes and half-life data for relevant compartments can be used to adjust the concentrations in the water and sediment.

For bath treatments in net pens or cages, the initial refinement would be to consider the total number of net pens and the total volume for dilution (note that this may not be allowed, depending upon local regulation). For bath treatments in net pens, hydrodynamic effects will predominate in the determination of the $PEC_{sw-refined}$. SEPA has developed approaches for calculating concentrations of medicines in net pen and cage systems (SEPA 2003). These approaches are suggested for use in the exposure evaluation, and they are briefly described below. The SEPA procedures provide for a short-term assessment tool and a longer term tool to evaluate bath treatments. The short-term evaluation uses a simple model, primarily governed by mean current speed at the site, distance to shore, and cage volume during treatment. The model assumes that the "chemical patch" formed by treatment is transported longitudinally at the mean current speed while spreading laterally at a rate determined by a dispersion coefficient. The time since release determines the area of the patch, and, assuming a constant depth, the volume of the patch can be calculated. The mean concentration of the medicine within the patch at the end of the evaluation period is thus the initial mass released divided by the volume. The predicted concentrations are suitable for comparison to data on short-term effects. For instance, in the SEPA procedures, 3-hour or 72-hour standards are used.

The SEPA procedures manual also provides a longer term model for medicines that have a longer residence time in the environment, due either to their inherent properties or to multiple applications. According to the manual, the longer term model is relevant for chemicals that are still present at potentially toxic concentrations after 72 hours. This model includes site-specific topographic and hydrodynamic parameters and allows the particulars of the treatment program to be specified, such as total area of cage group, the number of cages, the depth of cages during treatment, the number of discrete treatments, and the interval between treatments. The long-term model simulates the dispersing plumes from each discrete bath treatment and predicts the path and concentration of these plumes throughout the period under assessment, during which time the concentration is reduced according to the specified decay half-life. Details of the assumptions of both models, the calculations used, and examples are provided in Annex G of the SEPA procedures manual (SEPA 2003).

For medicated feeds used in net pens and cages, the leaching (or lack thereof) of the medicine from the uneaten food and feces can be important for the analysis of fate and transport. Factors such as settling velocities, resuspension, bioturbation, local currents, pore water exchange, and burial rate will affect the concentration of the medicine that ends up in the sediment versus the water column. Data on adsorption and desorption (e.g., OECD 106) could be used to evaluate the potential for leaching of the medicine out of the accumulated feed and feces below the cage. Data on the rate of biodegradation could be used to refine the PEC further. The SEPA guidance document discusses methods for predicting the residual concentrations of 2 specific antiparasitic chemicals and the benthic impacts of these residues. Site-specific characteristics (e.g., hydrography, depth, size of fish farm, cage positions, shape, size and orientation of cages, and bathymetry) as well as chemical-specific data (feeding rate, percent consumed, etc.) are used in a complex model to predict the "deposition footprint" of the 2 specific chemicals for which the model has been developed. Details of the model are provided in Annex H of the guidance manual (SEPA 2003).

When a bath treatment is used in a flow-through system, it may be done under either static or flow-through conditions. Factors that can be considered to refine the PEC_{sw} include the number of raceways being treated, the presence or absence of wastewater treatment facilities at the facility (such as settling ponds), and the dilution afforded upon discharge from the hatchery. For flow-through systems, historical discharge and receiving water data are needed to calculate the amount of dilution in the environment. A worst-case estimate can be derived using a high flow rate for the hatchery discharge and a low flow rate for the receiving stream. Prior to discharge, fate processes for the medicine may or may not be important for refining the $PEC_{initial}$, depending upon the duration of the treatment period for the medicine, the size of the facility, the flow rates in the facility, and other factors. In the receiving water the rates of various fate processes (e.g., hydrolysis, photolysis), if known, may be used to refine the PEC_{sw}.

Similar considerations are used to refine the PEC_{sw} for medicated feeds in flow-through systems. Again, the flow conditions within the facility and in the receiving water are crucial parameters for developing worst-case and typical-case scenarios. Fate processes are potentially more important variables for scenarios involving medicated feeds than for bath treatments, because the duration of treatment may be considerably longer in the former case. For oxytetracycline used in an aquaculture facility with fixed hydrodynamic conditions, Rose and Pedersen (2005) observed that the most important fate processes influencing the oxytetracycline concentrations in the receiving water were the settling pond biosolids' (i.e., fish feces) load, the biosolids' settling velocity, leaching from biosolids, and the distribution coefficients for oxytetracycline bound to particles in the receiving water. Sediment water fate studies (e.g., OECD 308) may be used to estimate the rates of degradation of the medicine within the hatchery and within receiving waters, and to make refinements to the PEC values.

4.3.2.3.11   PEC for Sediment

Current VICH phase II guidance only recommends that sediment exposure be addressed if there is predicted toxicity in the water column (i.e., risk quotient for the aquatic invertebrate study is ≥ 1). For determining the $PEC_{sediment\text{-}initial}$, the VICH guidance document assumes that partitioning between sediment and water are complete and that the sediment and water are in equilibrium. However, monitoring data have shown that medicines such as emamectin benzoate can be detected in sediments below net pens (Haya et al. 2005), even when the concentrations in water are below detection limits. Therefore, the current VICH trigger value (i.e., aquatic invertebrate risk quotient ≥ 1) may not be appropriately conservative for determining the need for a sediment assessment.

A worst-case PEC$_{sediment}$ would assume that none of the medicine ends up in the water column and that it is all in the uneaten food and feces. The simplest method for this calculation would be to use the mass of the medicine that is fed to the fish, and assume that it all partitions into the sediment, assuming incorporation to a certain depth. To refine this value, it is important to account for the amount of the medicine metabolized by the fish versus how much ends up in feces, desorbs from the feces, and degrades, as well as various sediment parameters (i.e., depth and density of sediment). For flow-through and recirculation systems, the presence of settling ponds or some other type of solids removal system is important in reducing environmental loadings of medicines. The SEPA guidance document provides a model for the deposition of residues of veterinary medicines in sediment under net pens.

4.4   CONCLUSIONS

In recent years there have been significant advances in our understanding of the sources and fate of veterinary medicines in aquatic systems. Alongside this, detailed guidance has been developed on regulatory approaches for assessing aquatic exposures, and a range of exposure modeling approaches and scenarios have been developed. In this chapter, we have provided an overview of recent research on the fate of veterinary medicines in aquatic ecosystems, and we have used available data to evaluate many of the modeling approaches. In addition, a range of simple models for use in aquaculture assessment have been proposed. However, there are still a number of significant gaps in our knowledge. Some of the major areas of uncertainty are highlighted below.

The information provided in this chapter on comparisons between predicted and measured exposure concentrations is the first attempt to provide these comparison data. This exercise showed that the available methods for assessing aquatic exposures as a result of terrestrial applications of veterinary medicines generally provide conservative estimates of exposure concentrations, with some notable exceptions, such as strongly sorbed compounds. The predicted concentrations of strongly sorbing antibiotics such as tetracyclines in surface water and groundwater tend to be underestimated, as the models do not consider colloidal or particle-bound transport. Studies to investigate the mechanisms of transport of highly sorbing substances and subsequent model refinements are therefore warranted.

In the case of aquatic exposure assessments related to aquaculture facilities, the available assessment methods require further development. The simple algorithms proposed in this chapter will make a significant contribution to the development of exposure assessment methods for veterinary medicines used in aquaculture. However, more sophisticated exposure models are needed, especially in the case of intensive net pen aquaculture. Exposure scenarios for different aquaculture systems (pond, net pen, flow-through, etc.) for specific applications of medicines (bath versus feed) are needed. Operational data are also needed for the aquaculture facilities to refine the exposure scenarios (e.g., flow rates used, dilution factors, and number of treatments). Additional monitoring data are needed to examine the appropriateness of the aquaculture exposure scenarios for screening-level risk assessments.

Although a large body of data is now available on the transport of veterinary medicines into aquatic systems, with the exception of some aquaculture treatments much less information is available on the fate and dissipation of veterinary medicines in receiving waters. Research is required to improve our understanding of the relative importance of partitioning processes for medicines (water, feces, etc.), degradation processes, and other dissipation mechanisms in order to determine the most appropriate way to calculate PECs for aquatic systems. As inputs are likely to be intermittent or pulsed for some medicines (e.g., bath treatments), more consideration may also be given to approaches that link the temporal variability of aquatic exposures to effects, such as the use of time-weighted averages. The degradation processes in water or sediment may result in the formation of transformation products. The persistence and fate of these substances in surface water bodies may be very different from those of the parent compound. For example anhydro-tetracycline, a degradation product of tetracycline, is known to be more mobile than the parent compound. Current exposure assessment scenarios do not take into account the presence of metabolites or transformation products of veterinary medicines that could be biologically active.

Exposure assessments typically do not take into account ecosystem-level effects that occur as a result of multiple inputs of veterinary medicines. These scenarios are quite common, as intensive aquaculture and agricultural operations tend to be clustered in restricted geographical areas. Under these scenarios, inputs from multiple sources could be cumulative for exposures of aquatic organisms to waterborne contaminants. Exposure assessment methods are also not designed to assess mixtures of veterinary medicines. Assessments are typically conducted on single active ingredients as part of an approvals process for the marketing of veterinary medicine formulations. However, there is potential for mixtures of chemicals to impact aquatic organisms in an additive or greater than additive manner; especially when the veterinary medicines have similar mechanisms of action (e.g., antibiotics). These issues are particularly important when considering exposures to veterinary medicines that are marketed as mixtures, such as the potentiated sulfonamide antibiotics.

In terms of risk management, more work needs to be done to identify beneficial management practices (BMP) that can be used to mitigate exposures of aquatic organisms. So far there have been hardly any studies to evaluate the capacity of BMPs such as optimized tillage practices, and maintenance of buffer strips and riparian zones to reduce aquatic exposures from the terrestrial application of veterinary medicines. In the case of current-use pesticides, there is ample evidence that inputs into aquatic systems can be mitigated by use of these BMPs. Harman et al. (2004) used a predictive model to demonstrate that transport of atrazine from agricultural fields can be significantly reduced by construction of sediment ponds, grass buffer strips, and wetlands. Farm ditches can also be important control features for removing pesticides before they reach watersheds (Margoum et al. 2003). Blankenberg et al. (2006) showed that the construction of small wetlands in first- and second-order streams that drain agricultural land can significantly reduce loadings to the watershed. In addition, BMPs used in aquaculture, such as periodic moving of net pens, feeding regimes for medicated feeds, and static versus flow-through bath treatments, have not been assessed for their efficacy in reducing aquatic exposures to veterinary medicines.

Overall, the development of exposure assessment methods for veterinary medicines is an ongoing process that will require continuous refinement. This chapter makes a significant contribution to this incremental process by identifying research priorities and making recommendations for regulatory approaches.

REFERENCES

Aga DS, Goldfish R, Kulshrestha P. 2003. Application of ELISA in determining the fate of tetracyclines in land-applied livestock wastes. Analyst 128:658–662.

Alderman DJ, Hastings TS. 1998. Antibiotic use in aquaculture. Intl J Food Sci Tech 33:139–155.

Armstrong A, Philips K. 1998. A strategic review of sheep dipping. Environment Agency R&D Technical Report P170. Environment Agency, Bristol, England.

Armstrong SM, Hargrave BT, Haya K. 2005. Antibiotic use in finfish aquaculture: modes of action, environmental fate, and microbial resistance. In: Hargrave BT, editor. Environmental effects of marine finfish aquaculture. Vol. 5 of The handbook of environmental chemistry. Berlin (Germany): Springer-Verlag, p 305–340.

Batt AL, Snow DD, Aga DS. 2006. Occurrence of sulphonamide antimicrobials in private water wells in Washington County, Idaho, USA. Chemosphere 64:1963–1971.

Bebak-Williams J, Bullock G, Carson MC. 2002. Oxytetracycline residues in freshwater recirculating systems. Aquaculture 205:221–230.

Benbrook CM. 2002. Antibiotic drug use in US aquaculture. IATP Report, February. Minneapolis (MN): IATP.

Björklund HV, Bondestam J, Bylund G. 1990. Residues of oxytetracycline in wild fish and sediments from fish farms. Aquaculture 86:359–367.

Björklund HV, Råbergh CMI, Bylund G. 1991. Residues of oxolinic acid and oxytetracycline in fish and sediments from fish farms. Aquaculture 97:85–96.

Blackwell PA, Boxall ABA, Kay P, Noble H. 2005. Evaluation of a lower tier exposure assessment model for veterinary medicines. J Agric Food Chem 53(6):2192–2201.

Blackwell PA, Kay P, Boxall ABA. 2007. The dissipation and transport of veterinary antibiotics in a sandy loam soil. Chemosphere 62(2):292–299.

Blankenberg A-G, Braskerud B, Haarstad K. 2006. Pesticide retention in two constructed wetlands: treating non-point source pollution from agricultural runoff. Intl J Environ Anal Chem 86:225–231.

Boxall A, Fogg L, Baird D, Telfer T, Lewis C, Gravell A, Boucard T. 2006. Targeted monitoring study for veterinary medicines. Environment Agency R&D Technical Report. Bristol (UK): Environment Agency. 73 p.

Boxall ABA, Fogg LA, Kay P, Blackwell PA, Pemberton EJ, Croxford, A. 2004. Veterinary medicines in the environment. Rev Environ Contam Toxicol 180:1–91.

Boxall ABA, Fogg LA, Baird DJ, Lewis C, Telfer TC, Kolpin D, Gravell A, Pemberton E, Boucard T. 2005. Targeted monitoring study for veterinary medicines in the environment. Environment Agency Science Report SC030183/SR. Environment Agency, Bristol, England.

Boxall ABA, Kolpin D, Halling-Sørensen B, Tolls J. 2003. Are veterinary medicines causing environmental risks? Environ Sci Technol 37:286A–294A.

Burka JF, Hammell KL, Horsberg TE, Johnson GR, Rainnie DJ, Speare DJ. 1997. Drugs
used in salmonid aquaculture — a review. J Veter Pharmacol Therap 20:333–349.

Burkhard M, Stamm S, Waul C, Singer H, Muller S. 2005. Surface runoff and transport of
sulfonamide antibiotics on manured grassland. J Environ Qual 34:1363–1371.

Cabello FC. 2006. Heavy use of prophylactic antibiotics in aquaculture: a growing problem for human and animal health and for the environment. Environ Microbiol
8:1137–1144.

Capone DG, Weston DP, Miller V, Shoemaker C. 1996. Antibacterial residues in marine
sediments and invertebrates following chemotherapy in aquaculture. Aquaculture
145:55–75.

Carlson, JC, Mabury, SA. 2006. Dissipation kinetics and mobility of chlortetracycline,
tylosin, and monensin in an agricultural soil in Northumberland County, Ontario,
Canada. Environ Toxicol Chem 25:1–10.

Chen S, Summerfelt S, Losordo T, Malone R. 2002. Recirculating systems, effluents, and
treatments. In: Tomasso JR, editor. Aquaculture and the environment in the United
States. Baton Rouge (LA): US Aquaculture Society, p 119–140.

Coyne R, Hiney M, O'Connor B, Kerry J, Cazabon D, Smith P. 1994. Concentration and
persistence of oxytetracycline in sediments under a marine salmon farm. Aquaculture 123:31–42.

[CVMP] Committee for Medicinal Products for Veterinary Use. 2006. Guideline on
environmental impact assessment for veterinary medicinal products, in support of
the VICH guidelines GL6 and GL38. London (UK): European Medicines Agency
(EMEA) Veterinary Medicines and Inspection, 59 p.

Daughton CG, Ternes TA. 1999. Pharmaceuticals and personal care products in the environment: agents of subtle change? Special report. Environ Health Persp Suppl.
107:907–938.

Davies IM, Gillibrand PA, McHenry JG, Rae GH. 1998. Environmental risk of ivermectin
to sediment-dwelling organisms. Aquaculture 163:29–46.

Dietze JE., Scribner EA, Meyer MT, Kolpin DW. 2005. Occurrence of antibiotics in water
from 13 fish hatcheries, 2001–2003 Intern J Environ Anal Chem 85:1141–1152.

FOCUS. n.d. Surface water. http://viso.ei.jrc.it/focus/sw/index.html.

Gaikowski MP, Larson WJ, Steur JJ, Gingerich WH. 2004. Validation of two models
to predict chloramine-T concentrations in aquaculture facility effluent. Aquacult
Engin 30:127–140.

Gupta S, Kumar K, Thompson A, Thoma D. 2003. Antibiotic losses in runoff and drainage from manure applied fields. USGS-WRRI 104G National Grant. Washington
(DC): US Geological Survey.

Halling-Sørensen B, Jacobsen A-M, Jensen J, Sengelov G, Vaclavik E, Ingerslev F. 2005.
Dissipation and effects of chlortetracycline and tylosin in two agricultural soils: a
field-scale study in Southern Denmark. Environ Toxicol Chem 24:802–810.

Halling-Sørensen B, Jensen J, Tjørnelund, J, Montforts MHMM. 2001. Worst-case estimations of predicted environmental soil concentrations (PEC) of selected veterinary
antibiotics and residues used in Danish agriculture. In: Kümmerer K, editor. Pharmaceuticals in the environment: sources, fate, effects and risks. Berlin (Germany):
Springer-Verlag, p 72–83.

Hamscher G, Abu-Quare A, Sczesny S, Höper H, Nau H. 2000a. Determination of tet-racyclines and tylosin in soil and water samples from agricultural areas in lower Saxony. In: van Ginkel LA, Ruiter A, editors. Proceedings of the Euroresidue IV conference, Veldhoven, The Netherlands, 8–10 May. Bilthoven (The Netherlands): National Institute of Public Health and the Environment (RIVM).

Hamscher G, Sczesny S, Abu-Quare A, Höper H, Nau H. 2000b. Substances with phar-macological effects including hormonally active substances in the environment: identification of tetracyclines in soil fertilised with animal slurry. Deutsch Tierärztl Wschr 107:293–348.

Hamscher G, Pawelzick HT, Hoper H, Nau H. 2005. Different behavior of tetracyclines and sulfonamides in sandy soils after repeated fertilisation with liquid manure. Environ Toxicol Chem 24:861–868.

Hargreaves JA, Boyd CE, Tucker CS. 2002. Water budgets for aquaculture production. In: Tomasso JR, editor. Aquaculture and the environment in the United States. Baton Rouge (LA): US Aquaculture Society, p 9–34.

Harman WL, Wang E, Williams JR. 2004. Reducing atrazine losses: water quality impli-cations of alternative runoff control practices. J Environ Qual 33:7–12.

Haya K, Burridge E, Davies IM, Ervik A. 2005. A review and assessment of environmental risk of chemicals used for the treatment of sea lice infestations of cultured salmon. In: Hargrave BT, editor. Environmental effects of marine finfish aquaculture. Vol. 5 of The handbook of environmental chemistry. Berlin (Germany): Springer-Verlag, p 340–356.

Hirsch R, Ternes T, Haberer K, Kratz KL. 1999. Occurrence of antibiotics in the aquatic environment. Sci Total Environ 225:109–118.

Huet M. 1994. Textbook of fish culture, breeding and cultivation of fish. 2nd ed. Oxford (UK): Fishing News.

Jørgensen SE, Halling-Sørensen B. 2000. Drugs in the environment. Chemosphere 40:691–699.

Kay P, Blackwell P, Boxall A. 2004. Fate and transport of veterinary antibiotics in drained clay soils. Environ Toxicol Chem 23:1136–1144.

Kay P, Blackwell PA, Boxall ABA. 2004. Fate of veterinary antibiotics in a macroporous tile drained clay soils. Environ Toxciol Chem 23(5):1136–1144.

Kay P, Blackwell PA, Boxall ABA. 2005. Column studies to investigate the fate of vet-erinary antibiotics in clay soils following slurry application to agricultural land. Chemosphere 60:497–507.

Kay P, Blackwell PA, Boxall ABA. 2005. Transport of veterinary antibiotics in overland flow following the application of slurry to arable land. Chemosphere 59:951–959.

Kreuzig R, Holtge S. 2005. Investigations on the fate of sulfadiazine in manured soil: laboratory experiments and test plot studies. Environ Toxicol Chem 24:771–776.

Kreuzig R, Holtge S, Brunotee J, Berenzen N, Wogram J, Schulz R. 2005. Test plot stud-ies on runoff of sulfonamides from manured soils after sprinkler irrigation. Environ Toxciol Chem 24:777–781.

Lalumera GM, Calamari D, Galli P, Castiglioni S, Crosa G, Fanelli R. 2004. Preliminary investigation on the environmental occurrence and effects of antibiotics used in aquaculture in Italy. Chemosphere 54:661–668.

Lazur AM, Britt DC. 1997. Pond recirculating production systems. Southern Regional Aquaculture Center (SRAC) Publication No. 4. http://srac.tamu.edu.

Lissemore L, Hao CY, Yang P, Sibley PK, Mabury S, Solomon KR. 2006. An exposure assessment for selected pharmaceuticals within a watershed in southern Ontario. Chemosphere 64:717–729.

Littlejohn JW, Melvin AAL. 1991. Sheep-dips as a source of pollution of freshwaters: a study in Grampian Region. J Chartered Instit Water Environ Manag 5:21–27.

Loke ML, Ingerslev F, Halling-Sørensen B, Tjørnelund J. 2000. Stability of tylosin A in manure containing test systems determined by high performance liquid chromatography. Chemosphere 40:759–765.

Losordo TM, Masser MP, Rakocy J. 1999. Recirculating aquaculture tank production systems: a review of component options. Southern Regional Aquaculture Center (SRAC) Publication No. 4http://srac.tamu.edu.

Lunestad BT. 1992. Fate and effects of antibacterial agents in aquatic environments. Chemotherapy in Aquaculture: From Theory to Reality conference, Bureau International des Epizootices, Paris, France.

Mackie RI, Koike S, Krapac I, Chee-Sanford J, Maxwell S, Aminov RI. 2006. Tetracycline residues and tetracycline resistance genes in groundwater impacted by swine production facilities. Animal Biotechnology 17:157–176.

Margoum C, Gouy V, Laillet B, Dramais G. 2003. Retention of pesticides by farm ditches. Rev Sci Eau/J Water Sci 16:389–405.

Masser MP, Jensen JW. 1991. Calculating area and volume of ponds and tanks. Southern Regional Aquaculture Center (SRAC) Publication No. 1. http://srac.tamu.edu.

Mazik PM, Parker NC. 2001. Semicontrolled systems. In: Wedemeyer GA, editor. Fish hatchery management. 2nd ed. Bethesda (MD): American Fisheries Society, p 241–284.

Meyer MT, Bumgarner JE, Varns JL, Daughtridge JV, Thurman EM, Hostetler KA. 2000. Use of radioimmunoassay as a screen for antibiotics in confined animal feeding operations and confirmation by liquid chromatography/mass spectrometry. Sci Total Environ 248:181–187.

Meyer MT, Ferrell G, Bunzgamer JE, Cole D, Hutchins S, Krapac I, Johnson K, Kolpin D. 2003. Occurrence of antibiotics in swine confined feeding operations lagoon samples from multiple states 1998–2002: indicators of antibiotic use. In: Proceedings 3rd International Conference on Pharmaceuticals and Endocrine Disrupting Chemicals in Water. National Groundwater Association Meeting, Minneapolis, 19–21 March.

Montforts MHMM. 1999. Environmental risk assessment of veterinary medicinal products. Part 1: other than GMO-containing and immunological products. RIVM report 601300 001, N120. National Institute of Public Health and the Environment, Bilthoven, The Netherlands.

Nessel RJ, Wallace DH, Wehner TA, Tait WE, Gomez L. 1989. Environmental fate of ivermectin in a cattle feedlot. Chemosphere 18:1531–1541.

Pouliquen H, Le Bris H, Pinault L. 1992. Experimental study of the therapeutic application of oxytetracycline, its attenuation in sediment and sea water, and implication for farm culture of benthic organisms. Marine Ecol Prog Series 89:93–98.

Rose PE, Pedersen JA. 2005. Fate of oxytetracycline in streams receiving aquaculture discharges: model simulations. Environ Toxicol Chem 24:40–50.

Samuelsen OB. 1989. Degradation of oxytetracycline in seawater at two different temperatures and light intensities and the persistence of oxytetracycline in the sediment from a fish farm. Aquaculture 83:7–16.

Samuelsen OB, Lunestad BT, Husevåg B, Hølleland T, Ervik A. 1992a. Residues of oxolinic acid in wild fauna following medication in fish farms. Dis Aquatic Org 12:111–119.

Samuelsen OB, Torsvik V, Ervik A. 1992b. Long-range changes in oxytetracycline concentration and bacterial resistance towards oxytetracycline in a fish farm sediment after medication. Sci Total Environ 114:25–36.

[SEPA] Scottish Environmental Protection Agency. 1999. Enamectin benzoate: an environmental risk assessment. Report of the SEPA Fish Farm Advisory Group, 66/ Edinburgh: Scottish Environmental Protection Agency.

[SEPA] Scottish Environmental Protection Agency. 2003. Regulation and monitoring of marine cage fish farming in Scotland: a manual of procedures. http://www.sepa.org. uk/guidance/fishfarmmanual/manual.asp.

Shepherd J, Bromage N, editors. 1992. Intensive fish farming. Oxford (UK): Blackwell Science.

Smith P, Donlon J, Coyne R, Cazabon DJ. 1994. Fate of oxytetracycline in a fresh water fish farm: influence of effluent treatment systems. Aquaculture 120:319–325.

Sommer C, Steffansen B, Overgaard Nielsen B, Grønvold J, Kagn-Jensen KM, Brøchner Jespersen J, Springborg J, Nansen P. 1992. Ivermectin excreted in cattle dung after sub-cutaneous injection or pour-on treatment: concentrations and impact on dung fauna. Bull Entomol Res 82:257–264.

Spaepen KRI, Leemput LJJ, Wislocki PG, Verschueren C. 1997. A uniform procedure to estimate the predicted environmental concentration of the residues of veterinary medicines in soil. Environ Toxicol Chem 16:1977–1982.

Steeby J, Avery J. 2002. Construction of levee ponds for commercial catfish production. Southern Regional Aquaculture Center (SRAC) Publication No. 1http://srac.tamu.edu.

Stickney RR. 2002. Impacts of cage and net-pen culture on water quality and benthic communities. In: Tomasso JR, editor. Aquaculture and the environment in the United States. Baton Rouge (LA): US Aquaculture Society, p 105–118.

Stoob K, Singer HP, Mueller SR, Schwarzenbach RP, Stamm CH. 2007. Dissipation and transport of veterinary sulfonamide antibiotics after manure application to grassland in a small catchment. Environ Sci Technol 41(21):7349–7355.

Tucker CS, Boyd CE, Hargreaves JA. 2001. Characterization and management of effluents from warmwater aquaculture ponds. In: Tomasso JR, editor. Aquaculture and the environment in the United States. Baton Rouge (LA): US Aquaculture Society, p 35–76.

[USEPA] US Environmental Protection Agency. 2004. Technical development document for the final effluent limitations guidelines and new source performance standards for the concentrated aquatic animal production point source category. Revised August 20. EPA-821-R-04-0. Washington (DC): Office of Water, US Environmental Protection Agency.

Veterinary Medicines Directorate. N.d. Downloads. http://www.vmd.gov.uk/downloads.htm.

[VICH] International Cooperation on Harmonization of the Technical Requirements for Registration of Veterinary Medicinal Products. 2004. Environmental impact assessment (EIAs) for veterinary medicinal products (VMPs) — phase II, draft guidance. August. http://vich.eudra.org/pdf/10_2003/g138_st4.pdf.

Westers H. 2001. In: Wedemeyer GA, editor. Fish hatchery management. 2nd ed. Bethesda (MD): American Fisheries Society, p 31–90.

Westers H, Pratt KM. 1996. Rational design of hatcheries for intensive salmonids culture, based on metabolic considerations. Prog Fish-Cult 39:157–165.

Weston DP 1996. Environmental considerations in the use of antibacterial drugs in aquaculture. In: Baird DJ, Beveridge MCM, Kelly LA, Muir JF, editors. Aquaculture and water resource management. London (UK): Blackwell Science, p 140.

Whitis GN. 2002. Watershed fish production ponds: guide to site selection and construction. Southern Regional Aquaculture Center (SRAC) Publication No. 1. http://srac. tamu.edu.

Willoughby S. 1999. Salmonid farming technology. In: Manual of salmonids farming. Oxford (UK): Fishing News Books, p 123–157.

Wrzesinski C, Crouch L, Gaunt P, Holifield D, Bertrand N, Endris R. 2006. Florfenicol residue depletion in channel catfish, *Ictalurus punctatus* (Rafinesque). Aquaculture 253:309–316.

Xu W, Zhu X, Wang X, Deng L, Zhang G. 2006. Residues of enrofloxacin, furazolidone and their metabolites in Nile tilapia (*Oreochromis niloticus*). Aquaculture 254:1–8.

5 Assessing the Aquatic Hazards of Veterinary Medicines

Bryan W. Brooks, Gerald T. Ankley,
James F. Hobson, James M. Lazorchak,
Roger D. Meyerhoff, and Keith R. Solomon

5.1 INTRODUCTION

In recent years, there has been increasing awareness of the widespread distribution of low concentrations of veterinary medicine products and other pharmaceuticals in the aquatic environment. Although aquatic hazard for a select group of veterinary medicines has received previous study (e.g., aquaculture products and sheep dips), until very recently less information has been available in the published literature for other therapeutic groups (Halling-Sørensen 1999; Jørgensen and Halling-Sørensen 2000; Ingerslev and Halling-Sørensen 2001; Koschorreck et al. 2002; Boxall et al. 2003, 2004b). The majority of available aquatic ecotoxicity information for veterinary medicines was generated from short-term (e.g., 24 to 96-hour) bioassays to meet requirements for product registrations (Boxall et al. 2004b). Limited information is available for partial life cycle or life cycle exposure scenarios and on hazards in lentic systems and lotic systems, particularly in arid to semiarid regions (Brooks et al. 2006, 2007).

Although aquatic hazard information for veterinary medicines is largely limited to acute toxicity data, the various classes of veterinary medicines are generally known to have specific biological properties, which are selected during the drug development process. It is possible that such information may be leveraged to focus future research and the screening of the potential hazards these compounds present to specific groups of nontarget organisms. For example, Huggett et al. (2003) describe a theoretical model that may be used to estimate impacts of selected veterinary medicines to fish, based on pharmacological information from other vertebrates.

This chapter considers the utility and applications of existing techniques (e.g., standardized toxicity tests), developing approaches (e.g., ecotoxicogenomics), and technologies or methods that may be used in the future with the existing knowledge of physiochemical (e.g., log K_{ow}) and pharmacological properties (e.g., mode of action) to characterize potential impacts of veterinary medicines in aquatic systems.

The chapter includes a critical evaluation of the state of veterinary medicine aquatic hazard assessment, and a characterization of available information for veterinary medicine impacts in aquatic systems. Furthermore, we identify data gaps and regulatory uncertainties or deficiencies, and provide recommendations for research needs.

5.2   PROTECTION GOALS

When assessing the risk of a compound to the environment and selecting aquatic testing strategies, it is essential to have clear protection goals. The protection goals developed during a recent SETAC Pellston Workshop on Science for Assessing the Impact of Human Pharmaceuticals on Aquatic Ecosystems (Williams 2005) would appear to be appropriate for veterinary medicines. The previous workshop concluded,

> The key aspects of aquatic ecosystems that should be protected are 1) Ecosystem functionality and stability — including ecosystem primary productivity (based on algae and plants) and the key phyla of primary consumers (especially invertebrates) that are essential to the sustainability of aquatic food webs; 2) Biodiversity — especially the potential to affect populations of potentially endangered species, taking into account both local and regional contexts; and 3) Commercially and socially important species, including shellfish (crustaceans and mollusks), fish, and amphibian populations. Finally, it is important to recognize the importance of linkages between ecosystem components. If an ecosystem component (population or group of populations) is strongly linked to other components, effects on that component have greater potential to cause secondary effects elsewhere in that ecosystem.

In the following sections we discuss potential approaches that can be used for environmental assessment of veterinary medicines to help achieve these goals.

5.3   APPROACHES TO ASSESS EFFECTS OF VETERINARY MEDICINES

Aquatic toxicity studies may be used in a number of ways (Chapter 3): they may contribute to the development of a risk assessment for a new product (prospective assessment), they may be used for routine monitoring of aquatic ecosystems such as in ecopharmacovigilance programs (retrospective or compliance assessment), or they may be used to help identify the causes of an observed impact on an ecosystem using approaches such as toxicity identification evaluation (retrospective assessment). In the following sections we describe existing and novel aquatic toxicity testing approaches that are appropriate for veterinary medicines and that could be used for any one of these purposes.

5.3.1   CURRENT METHODS OF ASSESSING AQUATIC EFFECTS FOR RISK ASSESSMENT

To date most developments in the area of toxicity of veterinary medicines to aquatic organisms have focused on prospective risk assessment, and several guidelines are now available on how the aquatic hazard of a veterinary medicine to aquatic organisms should be assessed. The most influential of these guidelines are those developed by VICH and are discussed in more detail in Chapter 3. The approach is a 2-phase process, and during phase 2 aquatic hazard testing is performed using a tiered approach.

5.3.1.1   Lower Tier Approaches

During the VICH phase I process (described in Chapter 3), compounds do not require additional study if they have a Predicted Environmental Concentration (PEC) or Environmental Introduction Concentration (EIC) of <1 µg L^{-1} in aquatic systems, or <100 µg kg^{-1} in soil (International Cooperation on Harmonization of Technical Requirements for Registration of Veterinary Products [VICH] 2000). Exceptions to this are a few therapeutic groups. compounds used in aquaculture, and endo- and actoparasiticides. For example, veterinary medicines applied to companion animals are not considered important because the mass potentially entering the aquatic environment is considered too small to result in exposures of ecological significance.

When an assessment of a veterinary medicine does not stop at phase 1 of the VICH process, acute algal, daphnid, and juvenile fish toxicity studies are performed at tier A of the VICH phase II process to estimate EC50 and LC50 values (VICH 2004). Predicted no-effect concentrations (PNEC) are then estimated by applying an assessment factor of 100 to the algal data and 1000 to the daphnid and fish data. The PNECs are then compared to the predicted exposure concentrations (see Chapter 4 for derivation) to generate a hazard quotient (HQ). If the HQ is < 1, the assessment is terminated. If an HQ > 1 is identified, tier B toxicity tests are performed that can include algal, cladoceran, sediment invertebrate, and fish assays to consider standardized sublethal responses such as growth or reproduction (VICH 2004).

5.3.1.2   Higher Tier Testing

The tier B tests (Table 5.1) incorporate responses to chronic exposures that differ in terms of the life cycle of the test organisms and the organisms for which they are surrogates. Only some of these tests allow observations of effects on all aspects of the life cycle, including reproduction, and of these, some only assess 1 type of reproduction (Table 5.1). Assessment factors of 10 are applied to no observed effect concentrations (NOECs) generated from these tier B tests, and the HQ calculation is repeated.

If the HQ remains > 1, the specific hazard of the compound can be further assessed during a tier C process in countries such as the United States, or risk management regimes can be considered. These additional tests may be required to address specific questions and test hypotheses related to the likely effects of the veterinary medicines on nontarget aquatic organisms. Specific recommendations

TABLE 5.1

Tier B tests proposed by the International Cooperation on Harmonization of Technical Requirements for Registration of Veterinary Products (VICH)

Test organism	Test guideline	Comments
Freshwater algae growth inhibition	OECD 201	Includes several life cycles and would likely allow the observation of subtle effects on growth, development, and reproduction. However, the form of reproduction may not include sexual modes.
Freshwater *Daphnia magna* reproduction	OECD 211	Includes 1 life cycle and would likely allow the observation of subtle effects on growth, development, and reproduction. However, the form of reproduction does not include sexual modes. ·
Freshwater fish, early life stage	OECD 210	A developmental bioassay that includes early stages of development and components of sexual differentiation, but not reproduction.
Freshwater sediment invertebrate species toxicity	OECD 218 and OECD 219	Includes survival and growth, but not reproduction.
Saltwater algae growth inhibition	ISO 10253	Includes several life cycles and would likely allow the observation of subtle effects on growth, development, and reproduction. However, the form of reproduction may not include sexual modes.
Saltwater crustacean chronic toxicity or reproduction	NA	Not specified but would include 1 life cycle and would likely allow the observation of subtle effects on growth, development, and reproduction. Sexual reproduction would likely be observed if the correct species is selected.
Saltwater fish chronic toxicity	NA	Not specified but could include reproduction.
Saltwater sediment invertebrate species toxicity	NA	Not specified but could include reproduction.

Note: NA = Not specified at this time.

Source: International Cooperation on Harmonization of Technical Requirements for Registration of Veterinary Products (VICH 2004).

for this testing are not currently included in VICH and FDA regulatory guidance documents. However, approaches described elsewhere in this chapter (e.g., tests and bioassays based on specific responses, such as hormonal activity) may be appropriate. For example, when the mode of action of a medicine in the target animal is known to be via hormone modulation, effects of reproductive function should be tested in an appropriate surrogate species such as fish. Some test protocols are available for this type of assessment (Ankley et al. 2001), and others are under development.

5.3.1.3 Limitations to Current Approaches

Single-species bioassays have greatly supported the improvement of water quality in many parts of the world. However, only relying on the endpoints employed in these standardized aquatic toxicity tests for prospective or retrospective contaminant decisions may not be sufficient (Cairns 1983), because these studies are not intended to predict structural or functional ecological responses to contaminants (Dickson et al. 1992; La Point and Waller 2000) and may not represent the most sensitive species responses (Cairns 1986). Furthermore, standardized test endpoints do not provide information on biochemical, developmental, behavioral, or transgenerational responses to veterinary medicine exposures.

Although assessment factors are applied in order to account for some of these issues, the assessment factors applied to toxicity results from tiers A and B have not been derived from empirical information for aquatic organisms exposed to veterinary medicines. This omission may have important implications for more sensitive species and ecologically relevant sublethal responses with high acute:chronic ratios (ACRs). For example, ACRs greater than 1000 have previously been reported for a number of pharmacologically active compounds in the environment (Huggett et al. 2002; Ankley et al. 2005; Crane et al. 2006).

In recent years, selection of appropriate measures of effect has been discussed for human medicines and personal care products and veterinary medicines (Daughton and Ternes 1999; Brooks et al. 2003; Crane et al. 2006). Because veterinary medicines are generally present in the environment at trace (ng L^{-1}) concentrations, traditional standardized ecotoxicity tests and endpoints may not be appropriate to characterize risk associated with aquatic exposures to certain compounds (Brooks et al. 2006). This problem is illustrated for 3 veterinary medicines with different modes of action (Table 5.2).

Diazinon is used in sheep dips as an organophosphorus insecticide to kill targeted terrestrial invertebrates species that are considered to be pests. Because *Daphnia magna* is an aquatic invertebrate species that is also sensitive to cholinesterase inhibition caused by compounds such as diazinon, a standardized toxicity test with this species using mortality and reproduction as the primary endpoints

---

TABLE 5.2

Example scenarios for veterinary medicines where aquatic hazards might or might not be found by current regulatory toxicity-testing approaches with standard endpoints

Compound	Bioassay organism	Hazard present	Hazard detected
Diazinon	*Daphnia*	Yes	Yes
Trenbolone	Juvenile fish	Yes	No
Oxytetracycline	Green algae	Yes	Possibly

will likely produce a sensitive measure of the potential hazard from this compound in surface waters. Oxytetracycline is a molecule that was selected to inhibit the growth of certain bacteria that result in disease conditions. There are no standard toxicity tests designed to assess the hazards of antibiotics to a community of microbes in surface waters. The combined results of studies with algae, especially blue-green algae, and soil microbes can provide an estimate of the potential hazards to aquatic microbes. So these standard tests may, or may not, allow an appropriate estimation of the hazard from an antibiotic in surface waters. The toxicity of the androgen trenbolone would not be appropriately characterized by the endpoints from an early life stage study with fish, a standard study conducted in tier B testing. Trenbolone can masculinize female fish (Ankley et al. 2003), but gender and reproduction are not determined in standard early life stage studies with fish.

To account for biological hazards associated with unique compounds like veterinary medicines, Ankley et al. (2005) recommended that test selection for a compound consider ecological attributes and appropriate species and endpoint relevance. Therefore, in the next section we review relevant approaches that may be used in conjunction with knowledge of the mode or mechanism of action of veterinary medicines to focus further investigations of their hazards to aquatic organisms and the development of postauthorization assessment methodologies.

5.3.2 Novel Approaches to Aquatic Effects Assessment

5.3.2.1 Use of Chemical Characteristics, Target Organism Efficacy Data, Toxicokinetic Data, and Mammalian Toxicology Data

Veterinary medicines are chemicals that are extensively evaluated for targeted efficacy, the safety of treated animals, and human safety. A significant number of studies are conducted to understand the physical, chemical, and structural characteristics of the molecules. Studies are also done to document the nature of the effects on the therapeutic target; the adsorption, distribution, metabolism, and excretion (ADME) of the chemical in the treated animal; and also potential unwanted toxicities in the treated animal. In order to protect humans from exposure to trace residues of the molecule in food sources, mammalian toxicology studies are conducted to characterize any reproductive or developmental effects, chronic whole organism and organ system toxicities, and cellular abnormalities. This information is interpreted by understanding the daily dose in the tested organisms, the resulting plasma exposure to the parent material, and the presence of metabolites. The basic environmental tests that are needed for the registered use of veterinary medicines also provide an important environmental hazard profile for these molecules.

The extensive testing of veterinary medicines for efficacy, safety of treated animals, and human safety may provide a significant amount of data that could be used to help identify information gaps in the environmental testing profile and to target appropriate testing to close these gaps. In the following sections, we describe potential applications for effects and bioaccumulation and bioconcentration assessment.

5.3.2.1.1   Effects Assessment

Information on the target mode of action of a veterinary medicine could potentially be used to select the most appropriate aquatic effect testing strategy (e.g., selection of the most appropriate test species and endpoints for use in ecological risk assessment and postauthorization monitoring). Examples of how the approach can be used are summarized in Table 5.3 and discussed in more detail below.

Treatments for microbial diseases such as antibiotics, antifungals, and anticoccidiostats are typically used to control specific types of microbes that can lead to respiratory, intestinal, and systemic infections as well as foot rot. The veterinary medicines developed for these purposes target the disease microorganisms by directly causing microbial cell death or by impeding the life cycle of targeted microbes through a variety of modes of action. Unless the veterinary product acts to improve the immune response in the host, the treatment does not achieve efficacy through a direct effect on the dosed animal. Understanding the modes of action for direct effects on disease microbes can help focus attention on possible

TABLE 5.3

Examples of how the results from mammalian tests can be used to target environmental effects testing

Target animal and mammalian results	Trigger for further evaluation	Taxa of interest	Endpoint of interest
Growth, development, and reproduction	Estrogen agonist activity	Fish	Development and reproduction
	PEC/Cmax at lowest result dose > 1 (especially when receptor mediated)	Fish	Development and reproduction, especially if receptor conserved in fish
Inhibition of cellular processes (e.g., ion transport or enzyme kinetics)	PEC/Cmax at lowest result > 1	Fish	Survival and growth, especially of cellular processes conserved
Thyroid effects	Hormone mediated	Frogs	Morphological transformation
Antibiotic efficacy	PEC/Efficacy Cmax >1	Similar microbial taxa or algae	Maximum inhibition concentrations, population growth
Ecto- and endoparasiticides	PEC/efficacy concentration > 1	Arthropods	Survival, growth
ADME kinetics slow with high K_{ow}	Long half-life and little metabolism	Fish, sediment, invertebrates	Possible significant bioaccumulation

hazards for related taxa in aquatic ecosystems with potential sensitivities to the same modes of action.

Veterinary medicines are also developed to have direct effects on parasites. These products can be delivered orally, topically, or by injection, but they are targeted at achieving a high enough exposure to kill or interrupt the life cycle of the parasite. Again, the treatment does not usually achieve efficacy through a direct effect on the dosed animal. Understanding the modes of action for direct effects on these parasites can provide the context for judging the adequacy of ecological hazard testing with invertebrate species.

Promotion of feed utilization efficiency and/or growth can be targeted through several very different modes of action. Some antimicrobials aim to modify the gut flora for more efficient digestion of feedstuffs and, therefore, better energy efficiency and growth for the treated host. Some antimicrobials target a reduction in the bacterial load in the host, resulting in less animal stress and better growth. A few veterinary medicines act through receptor-mediated modes of action to modify basal metabolism or augment hormonal action on growth. Known modes of action might be extrapolated to evaluate ecological hazards for similar aquatic taxa, or species with similar receptor-mediated physiological responses.

Treatment of veterinary medical conditions and aids for handling animals by veterinarians may act through a variety of modes of action. Some may be receptor mediated. Some might occur through direct modification of cellular processes; for example, through inhibition of enzyme kinetics or ion transporter activity. Other medicines may rely directly on the physical-chemical properties of the treatments (e.g., antifoaming agents for bloat). Again, knowledge of the mode of action targeted for these types of veterinary medicines can be useful in evaluating the types of hazards and species at risk when the chemical moves into aquatic ecosystems.

Mammalian toxicology studies can also reveal important clues to the potential effects of veterinary medicines in aquatic species. If developmental or reproductive effects occur at low doses, it may be important to evaluate further the potential for these effects to occur in fish. Chronic effects or unusual pathology noted in chronic mammalian studies could be used to identify important endpoints to evaluate in aquatic vertebrates. Tissue changes resulting from hormone-mediated effects could suggest sensitive species to test. For example, frogs might be tested for temporal changes in the transformation from tadpoles to air-breathing adults when a chemical is known to have thyroid receptor activity in mammals.

In order to assess the level of sensitivity at which these modes of action or toxicological endpoints occur, it is also important to relate the ADME characteristics of the veterinary medicines to their effects. The pharmacokinetic and toxicokinetic profiles of the molecules usually provide an understanding of the maximal plasma concentrations (C_{max}) and total exposure (area under the exposure curve) for the parent material and the primary metabolites to help explain the pharmacodynamics and toxicodynamics of the treatment in mammals. The plasma concentrations also help explain the activity of antimicrobial agents in vivo in relationship to their activity in vitro. These exposure–effect relationships

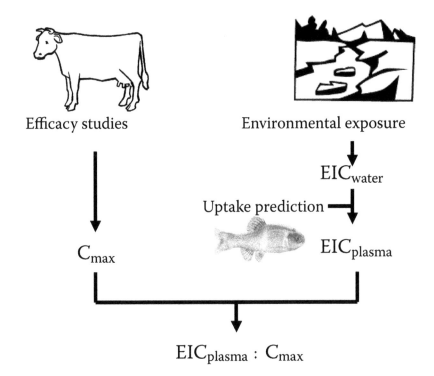

Efficacy studies   Environmental exposure

EIC_{water}

Uptake prediction

C_{max} EIC_{plasma}

$EIC_{plasma} : C_{max}$

FIGURE 5.1 Screening assessment approach to target aquatic effects testing with fish from water exposure. *Note*: EIC = environmental introduction concentration.

can be used to evaluate directly the potential for related effects on environmental species that are exposed to predicted environmental concentrations (PECs) calculated for the parent material (Huggett et al. 2003; and see Figure 5.1). If the predicted environmental concentrations of a veterinary medicine could cause concentrations in fish plasma levels that are near the C_{max} in mammalian plasma, resulting in efficacy or toxicity, it could be important to evaluate the endpoint further in a toxicity test with an appropriate environmental species.

5.3.2.1.2 Use of Chemical Characteristics and ADME Data in the Assessment of Bioaccumulation Potential

The ADME of a veterinary medicine in mammals can also, in conjunction with physical-chemical properties such as pK_a and solubility, provide some basis for estimating uptake and depuration characteristics in fish. Distribution within an aquatic vertebrate and the types of metabolism can parallel those found in mammals, although the kinetics and excretion routes in fish can be quite different. Absorption across the gut in mammals could lead to first-pass metabolism through the liver, whereas the route of exposure to somewhat soluble molecules is probably dominated in fish by absorption across the respiratory surfaces. Molecules that are found to be deeply distributed into the fatty tissue in mammals, or that are poorly metabolized and excreted, could also tend to accumulate in aquatic organisms and, perhaps, sediments. Extraordinary concentration of residues in particular tissues, such as reproductive organs, might also lead to concern for maternal deposition of active material in eggs for an environmental species. Other molecules that could appear to have the potential to bioaccumulate in fish, based on low solubility or high K_{ow}, might actually be as easily metabolized and excreted by fish as they are demonstrated to be through ADME studies in mammals. An illustration of the use of ADME data to design testing strategies for the aquaculture parasiticide emamectin benzoate is provided below.

Emamectin benzoate is synthetically derived from the natural product abamectin. Data have been developed for several applications in addition to aquaculture. Existing data include physicochemical properties, pharmacokinetics and metabolism data in fish and mammalian species, and bioaccumulation data for invertebrate species in the laboratory and field studies (Hobson 2004).

Physicochemical properties for emamectin benzoate are presented in Table 5.4. The vapor pressure, 4×10^{-3} mPa, indicates that the material is unlikely to enter or

TABLE 5.4
Physicochemical characteristics of emamectin benzoate

	4″-epimethylamino-4″-deoxyavermectin B1a benzoate (MAB1a)	4″-epimethylamino-4″-deoxyavermectin B1b benzoate (MAB 1b)
Composition (%)	> 90	< 10
Empirical formula	$C_{49}H_{75}NO_{13}C_7H_6O_2$	$C_{48}H_{73}NO_{13}C_7H_6O_2$
Molecular weight	1008.26	994.23
	Technical material	
Solubility (mg L^{-1})		
pH 5.03	320 ± 30	
pH 7.04	24 ± 2	
pH 9.05	0.1 ± 0.1	
Seawater	5.5	
Dissociation constant (pK_a)		
Benzoic acid group	4.2 ± 0.1	
Methylamino group	7.6 ± 0.1	
Log K_{ow}		
pH 5.07	3.0 ± 0.1	
pH 7.00	5.0 ± 0.2	
pH 9.04	5.9 ± 0.4	
Vapor pressure (mPa)	4×10^{-3}	
Melting point (°C)	141–146	
Density g cm^{-3}	1.20 ± 0.03	

persist as a vapor in the atmosphere. Solubility is pH dependent, ranging from 0.1 to 320 mg L^{-1}, and is 5.5 mg L^{-1} in seawater. It is reported to be soluble in chloroform, acetone, and methanol but insoluble in hexane. The pK_a values of 4.2 and 7.6 indicate that at the pH of seawater, the molecule will be in an ionized form, which may lead to the molecule binding to surfaces by ionic processes as well as partitioning due to the hydrophobic nature of the molecule. The log K_{ow} increases with increasing pH, with a value of 5.0 reported at pH 7. Although the hydrophobicity of the molecule may indicate a potential for bioaccumulation, the molecular weight (1008), the molecular size, and the polar characteristics of the molecule indicate it will not be lipophilic (i.e., will not preferentially bioconcentrate in fat) under biological conditions. The molecular size is large, which may limit absorption of this chemical. Although it has a moderately high K_{ow} (Table 5.4), it retains a measure of polarity. Both solubility and K_{ow} are pH dependent, indicating ionizable substituents. The polar nature of the molecule is reflected in the reported solubility of 5.5 mg L^{-1} in seawater (pH 7) and the observed solubility in the polar solvents acetone, chloroform, and methanol, contrasted with insolubility in hexane (a nonpolar, lipophilic solvent).

These characteristics indicate that emamectin may not appreciably biocentrate or biomagnify in aquatic organisms relative to many historical contaminants with log K_{ow} values > 5. This is supported by the observation that radiolabeled emamectin benzoate is not preferentially distributed to fat by either oral or intravenous administration. In salmon, rats, and goats, emamectin benzoate residues were found in a range of tissues including muscle, liver, and kidney at concentrations of the same order of magnitude as fat, and it appeared to depurate from fat at a similar rate as other tissues (Hobson 2004).

Biologically, emamectin benzoate does not demonstrate marked bioaccumulation in fish or invertebrates in the laboratory or in field studies. The highest rate of accumulation of residues observed in biota occurs with dietary exposure, but the highest bioaccumulation factors (BAFs) are observed in organisms exposed in water. Whole-body and tissue residues and pharmacokinetic studies show that emamectin benzoate is readily absorbed and metabolized and is excreted as parent and metabolites in fish and invertebrate species. Although excretion is somewhat prolonged in fish due to enterohepatic circulation, BAFs are consistently low (ranging from 9×10^{-5} to 116). Depuration is rapid when exposure to emamectin benzoate is removed. Sustained accumulation of emamectin benzoate or the desmethylamino metabolite was not observed in filter-feeding organisms (e.g., mussels) outside the footprint of the net pen in field studies (Hobson et al. 2004; Telfer et al. 2006).

In summary, despite a relatively high log K_{ow} of > 5.0 at environmentally relevant pH (pH 7) and moderate biological persistence in fish, when the existing data, including ADME, are considered bioconcentration in fish can be projected to be low. Moderate biological persistence of accumulated compounds in fish is related to retention of residues in vertebrates via enterohepatic circulation following substantial dietary exposure.

5.3.2.2   Use of Ecotoxicogenomics in Ecological Effects Assessment

As described above, identification of the mode of action (MOA) of veterinary medicines through prior knowledge serves as a logical basis for test and endpoint selection. However, an uncertainty associated with this is the possibility that a given test chemical could cause toxicity through multiple pathways and, as such, might produce unexpected impacts in nontarget species. In other regulatory programs with chemicals for which a priori MOA information is available (e.g., pesticide registration), uncertainty concerning multiple MOAs historically has been addressed through the routine collection of a large amount of data from several species and experimental designs. Collection of sometimes unused data in this fashion is not an efficient use of resources. Emerging techniques in the field of genomics have promise with respect to addressing MOA uncertainties in a more resource-effective manner. Specifically, in the case of veterinary medicines, it is conceptually reasonable that genomic data could be used to ascertain whether a chemical could cause toxicity through an unanticipated MOA. Below we describe in greater detail how this may be achieved.

The past few years have produced an explosion of analytical and associated bioinformatic tools that enable the simultaneous collection of large amounts of molecular and biochemical data indicative of the physiological status of organisms from bacteria to humans (MacGregor 2003; Waters and Fostel 2004). These tools, broadly referred to as genomic techniques, enable the collection of "global" information for an organism concerning gene or protein expression (transcriptomics and proteomics, respectively) or endogenous metabolite profiles (metabolomics). The amount of biological information that can be derived via genomic techniques is immense; for example, in humans it is estimated that the transcriptome, proteome, and metabolome include, respectively, 30 000, 100 000, and more than 2000 discrete elements (Schmidt 2004). This type of information has many potential uses, but one that is especially promising to the field of toxicology is identification of toxic MOAs. Specifically, it has been proposed that genomic (or, more precisely, toxicogenomic; Nuwaysir et al. 1999) data can serve as the basis for defining and understanding toxic MOAs in bacteria, plants, or animals exposed to chemical stressors. Specifically, changes in gene, protein, and/or metabolite expression can be highly indicative of discrete toxic MOAs. There has been a significant amount of work relative to the use of toxicogenomics to delineate MOAs in studies focused on human health risk assessment, and, although comparable work in the field of ecotoxicology initially lagged behind, there recently has been a steady increase in the development and application of toxicogenomic approaches in species and situations relevant to ecological risk assessment (Ankley et al. 2006).

There are different advantages (and challenges) in conducting transcriptomic versus proteomic versus metabolomic studies; an analysis of these is beyond the scope of this chapter. However, all the approaches can be useful for delineating toxic MOAs. To date, the most common approach to defining MOAs has been through the evaluation of changes in the transcriptome via high-density microarrays (or gene chips). Microarrays for different rat and mouse strains (typically representing several thousand genes) have been used fairly extensively for MOA-oriented toxicology studies over the past few years. Recently, the genomic information needed to develop comparable arrays (in terms of numbers of gene products) has become available for a number of species relevant to regulatory ecotoxicology, including *Daphnia* sp., rainbow trout, and the fathead minnow (Lettieri 2006). What differs between available mammalian arrays and those that have been (or will be) developed for invertebrates and fish is the degree of annotation (identification) of gene products present on the gene chips. Hence, in these nonmammalian models, it can be difficult (due to a lack of information concerning the complete genome or DNA sequence) to know precisely what genes are changing in a microarray experiment. However, this data gap does not necessarily preclude using alterations in gene expression as the basis for identification of toxic MOAs. Specifically, it is possible to assign MOAs to test chemicals with currently unknown MOAs through consideration of changes in overall patterns of gene expression and comparison of these patterns to those generated using chemicals with established MOAs. This approach, termed profiling (or "fingerprinting"), enables the application of toxicogenomic techniques to species for which the entire genome has not been sequenced. The above discussion, although focused on gene response (transcriptomic data), can analogously be extended to the collection and use of proteomic and metabolomic data for defining toxic MOAs.

There are several points in the veterinary medicine testing process where toxicogenomic data could potentially be useful. As alluded to above, one important use would be to identify where a chemical possesses MOA(s) different from (or, more typically, in addition to) that which is anticipated. The most straightforward approach to achieve this would be to conduct the base tests used in tier 1 of the risk assessment process (i.e., short-term assays with algae, cladocerans, and fish) with a set of reference compounds with well-defined toxic MOAs to develop a "library" of profile or fingerprint data. The reference chemicals should encompass toxicity pathways expected to occur in the various classes of veterinary medicines that may be tested (Ankley et al. 2005). For example, for model estrogenic and androgenic hormones, estradiol and trenbolone, respectively, would be logical reference materials, whereas a reference organophosphate ectoparasiticide might be diazinon. Once molecular profile data have been assembled for reference chemicals for the base test species, it should then be possible to compare fingerprints generated for a new "unknown" veterinary medicine to their expected profile (based on a priori MOAs). Congruence with the expected profile would provide strong direct evidence that the chemical does not possess an unanticipated MOA and that the suite of tests selected for the assessment is suitable for the task. If, however, molecular response profiles differ from what is expected, this could be taken as evidence that the chemical may act via additional toxicity pathways that the test suite might not adequately capture. In this scenario, the pattern of responses may be indicative of another MOA (reference chemical) present in the reference chemical library, or it may differ completely from previously generated fingerprints.

In either case, testing in addition to baseline assays (e.g., alternate or additional species, longer durations, or additional endpoints) might be warranted.

Finally, toxicogenomic data collected in conjunction with the base assay testing of reference chemicals would logically serve as the basis for the selection of discrete gene, protein, or metabolite biomarkers suitable for field studies and broad-scale monitoring studies with complex mixtures of veterinary medicines (and other chemicals). Molecular responses (biomarkers) unique to specific toxicity pathways would be extremely valuable to diagnostic and retrospective studies at the watershed scale.

5.4   APPLICATION FACTORS AND SPECIES SENSITIVITIES

During the risk assessment process for veterinary medicines, application, or assessment factors (AFs) are applied to the ecotoxicity test results to derive PNECs. These are ultimately used in the derivation of PEC/PNEC hazard quotients commonly used in deterministic assessments. The AF is a safety or extrapolation factor applied to an observed or estimated concentration to arrive at an exposure level that would be considered safe. Historically, in mammalian toxicology AFs sometimes called safety factors are used to account for extrapolation from laboratory animals to humans, or extrapolation from acute to chronic data (Cassarett et al. 1986). In ecotoxicology, AFs are used to account for unknown variability such as interspecies, intraspecies, or acute-to-chronic extrapolation when only a single data point or a limited data set is available (one or a few acute toxicity values). Generally a factor of 10 is applied to account for each area of expected variability, though empirical support of such a factor in ecotoxicology is not transparently communicated in existing regulatory documents.

In the evaluation of tier 1 results from VICH, a single very low toxicity value (acute or chronic) may represent a very sensitive species relative to the range of toxicity values for other species, or this toxicity value may be indicative of a sensitive taxonomic group such as algae. Generation of toxicity data for a wider range of species may be justified to evaluate adequately the hazard of a veterinary medicine and potentially to improve the characterization of hazard in risk assessment. With additional data, the use of AFs may be replaced or incorporated into more sophisticated analyses.

An analysis of the lowest toxicity value can be made in the context of the entire aquatic toxicology database. This can indicate what species or taxonomic groups should be tested and, when results are available, can show how the additional data contribute, or not, to an improved characterization of hazard. Toxicity data for an antibiotic used in aquaculture are presented here to illustrate such an analysis (Table 5.5). In this example, toxicity data were initially reported for a single algal species (*Skeletonema costatum*). This data point is substantially more sensitive when compared to the range of other species tested.

An alternative to using AFs is to utilize species sensitivity distributions (SSDs), a probabilistic analysis of hazard data. In this approach, the toxicity

TABLE 5.5
Predicted no-effect concentrations (PNECs) for aquatic organisms exposed to an antibiotic

Organism	EC50/LC50 (mg L⁻¹)	NOEC (mg L⁻¹)	Application factor	PNEC (mg L⁻¹)
Oncorhynchus mykiss	> 780	780	250	3.12[a]
Lepomis macrochirus	> 830	830	250	3.32[a]
Daphnia magna	> 330	< 100	250	1.32[a]
Litopenaeus vannamei	> 64	4	50	0.08[b]
Navicula pelliculosa	61	< 0.493	10	0.0493[c]
Pseudokirchneriella subcapitata	1.5	0.75	10	0.075[c]
Skeletonema costatum	0.0128	0.0042	10	0.00042[c]
Bacillus subtilis	0.4[d]		10	0.04[c,e]

[a] An application factor of 250 was used to account for interspecies and intraspecies variation and extrapolation from acute to chronic data.

[b] An application factor of 50 was used to account for intraspecies variation in this species, and a factor of 5 is added to account for use of early life stage data.

[c] These values already represent chronic endpoints.

[d] Maximum inhibition concentration (MIC).

[e] This MIC value was adjusted by a factor of 10 to account for intraspecies variation in calculation of the PNEC.

data for aquatic organisms are compared by graphing the concentration of exposures for various toxicity endpoints (on a log scale) on the x-axis for individual species. These values are graphed against the probability scale on the y-axis (Figure 5.2). This provides a normal distribution of the sensitivities for species tested. This distribution is assumed to be representative of the normal distribution of all species that might potentially be exposed to a compound. This approach to characterizing hazard using SSDs has been used in presenting hazard data and risk characterization in the regulation of pesticides and development of water quality criteria in the United States and for deriving PNECs, environmental risk limits (ERLs), and ecotoxicological soil quality criteria (ESQC) in Europe (Suter 2002; Solomon and Takacs 2002; SETAC 1994).

SSDs are applied in a more qualitative sense in evaluating the data for this antibiotic. *S. costatum* data are presented as the lower tail of a larger distribution of sensitivities for aquatic and marine species exposed to the antibiotic. Figure 5.2 illustrates the relative sensitivities of animal species to aquatic exposure. This figure is a plot and regression of acute toxicity values (•) (e.g., acute LC50 or EC$_{50}$ values) and NOEC concentrations (○) by log concentration on the x-axis and probability distribution on the y-axis. The latter is a ranked distribution of sensitivities (toxicity benchmark values) using a probability scale. This scale on the y-axis is a linear transformation of the sigmoid normal distribution, similar to a probit

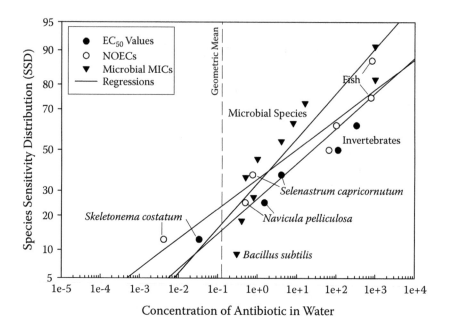

FIGURE 5.2 Species sensitivity distributions for aquatic organisms exposed to an antibiotic in water.

scale used in presenting mortality data. The toxicity values are ranked and evenly distributed across the probability scale and are assumed to represent the normal distribution of the toxicological response of aquatic organisms to a chemical in water.

As can be seen in Figure 5.2, there is a gradient of sensitivities for both acute (•) and chronic (○) NOEC distributions with the fish being the least sensitive and the algal species being the most sensitive. In addition to data for *S. costatum*, data points for 2 additional algal species are included and indicate that *S. costatum* is the most sensitive species of alga. Additional data in this case contribute to a refined characterization of algal sensitivities and to the relationship of algal sensitivities to the broader range of species and taxonomic groups.

A cursory evaluation of the applicability of the AFs used in the VICH approach can be made by comparison of the range of concentrations for species sensitivity values. For example, sensitivities of microbial (prokaryotic) species (▲) show a range of toxicological responses over more than 3 orders of magnitude and overlap with the eukaryotic species (fish, invertebrates, and algae). The most sensitive microbial species, *B. subtilis*, is protective of > 90% of all microbial species potentially exposed to the antibiotic, based on the SSD (Figure 5.2). A large AF would therefore not be needed in evaluating the risk to microbial species. This chronic data and interspecies variation is explained in the distribution of the maximum inhibition concentrations (MIC) for 10 species in the SSD. In this case an AF of 10 is used, conservatively, to account for intraspecies variability and calculate the PNEC for microbes in this example.

If only 1 microbial species was tested and if NOECs for algae are considered to be representative of chronic data, the *S. costatum* NOEC is nearly 2 orders of magnitude lower than those of the additional algal species, *N. pelliculosa* and *S. capricornutum*. Application of a hundredfold AF to the geometric mean of the 3 algal species tested would still protect *S. costatum*, the most sensitive algal species tested in this example.

Based on the EC50 and NOEC values for *S. costatum*, as shown in Figure 5.2, this species is the most sensitive in the distribution of measured biological responses and could have contributed to an overly conservative characterization of hazard. The addition of toxicity data for more species and analysis of the additional data in the context of the overall toxicity data set using the SSD is an example of the type of analytical tool that, when needed, can complement current regulatory processes and contribute to a refined characterization of hazard. For example, Brain et al. (2006) recently demonstrated the use of SSDs and the conceptually similar probabilistic ecological hazard assessment (PEHA) approaches for chemical toxicity distributions and intraspecies endpoint sensitivity distributions to characterize antibiotic effects on aquatic higher plants. Use of such PEHA approaches as part of the hazard and risk characterization processes allows for a probabilistic evaluation of toxicity data, which may reduce uncertainty when compared to deterministic approaches using default AFs.

5.5 EFFECTS OF VETERINARY MEDICINES IN THE NATURAL ENVIRONMENT

So far, we have discussed the different laboratory-based approaches that can be used to assess the effects of veterinary products on the aquatic environment and described methods for analyzing these data. However, aquatic ecosystems are much more complex. For example, 1) exposure to veterinary medicines may be episodic in nature; 2) due to matrix effects, the bioavailability in the environment may be very different than in the laboratory; 3) the parent medicine may be metabolized in the treated animal or be degraded in the environment, and it may be the metabolites or degradates that pose the risk; 4) veterinary medicines are highly unlikely to exist in the environment alone but will be present alongside other medicines as well as other aquatic contaminants, such as human medicines, pesticides, and nutrients; 5) veterinary medicines may be distributed as racemic mixtures, which may result in enantiospecific fate, exposure, and toxicity; and 6) aquatic ecosystems comprise interlinked communities of organisms, and exposure to veterinary medicines may result in indirect effects on a taxon that would never be picked up in a single-species laboratory study. In the following sections we discuss these different issues and, where possible, provide recommendations on how they can be considered in the hazard assessment process.

5.5.1 EPISODIC EXPOSURES

Laboratory toxicity studies are generally performed at fixed concentrations and for fixed durations. However, pulsed exposures will be particularly relevant for veterinary medicines associated with intensive livestock facilities (such as concentrated animal-feeding operations [CAFOs or feed lots]) and aquaculture, as discussed in Chapter 4. Transport of veterinary medicines from CAFOs and land receiving manure and slurry can be anticipated during rainfall events. Concentrations will then decline over time as both surface and groundwater flow recedes (e.g., Boxall et al. 2002a; Kay et al. 2004). With respect to aquaculture, several applications of a veterinary medicine can occur, and these can last from a few hours to days (see Chapter 4). Effects assessments should therefore take into account duration, magnitude, and frequency (e.g., pulsed or episodic) scenarios of exposures. To assess the implications of pulsed exposures, an approach can be used that is conceptually similar to that taken by the USEPA to protect aquatic systems with their water quality criteria (i.e., using a "criterion maximum concentration" [CMC] and a "criterion continuous concentration" [CCC]). The CMC is the acute concentration that cannot be exceeded for a 1-hour average once every 3 years on average, and the CCC is based on not exceeding a 4-day average once every 3 years on average. So, when assessing either prospective or retrospective ecological risks, veterinary medicines may not pose a risk if their predicted environmental concentrations do not exceed the predicted CMC or CCC values. However, traditional approaches for developing CMC or CCC values rely on toxicological benchmark responses from standardized toxicity tests, which may not be appropriate for some veterinary medicines. It may be possible to develop CMC or CCC values based on mechanistic responses or biomarkers of effect of ecological relevance using SSDs for taxa with toxicological targets specific to a veterinary medicine.

5.5.2 MATRIX EFFECTS

The impact of a veterinary medicine on aquatic organisms may depend on the form in which it enters the aquatic system and the properties of that receiving system, all of which may affect the bioavailability of a compound. Environmental fate studies (e.g., Kay et al. 2004) indicate that veterinary medicines used to treat livestock can enter the environment associated with colloids or suspended solids. Studies with other groups of compounds would indicate that the bioavailability, and hence effects, of these colloid and suspended solid-associated substances will be greatly reduced compared to the dissolved form of the medicine (see Boxall et al. 2002b for review). Sequestration of substances into sediments will also reduce exposures to water column organisms and generally also reduces toxicity to sediment-dwelling organisms that are exposed through pore water.

The disposition of a veterinary medicine in an aquatic ecosystem will depend on a range of mechanisms including hydrophobic partitioning, cation exchange, cation bridging, surface complexation, and hydrogen bonding (Tolls 2001). It is likely that many of these processes will also play a role in determining the bioavailability and uptake of a veterinary medicine in the aquatic environment. The relative importance of each mechanism will be determined by the characteristics of the aquatic system, including pH, hardness, dissolved organic carbon concentration, cation exchange capacity, and suspended solids concentration. Work on other groups of chemicals indicates that salinity may also be important in determining toxicity. Although modeling approaches are available for assessing the impacts of many of these variables on selected distribution mechanisms (e.g., Di Toro et al. 1991; Hoke et al. 1994; Santore et al. 2001; de Schamphelaere and Janssen 2002), it is not clear whether they can be applied to all classes of veterinary medicines, particularly for substances that are ionizable at environmentally relevant pH levels (Kümmerer et al. 2005).

5.5.3 METABOLITES AND DEGRADATES

Current approaches to the assessment of potential veterinary medicine effects in the environment generally do not consider metabolites (produced in the treated animal) or degradates (formed in the environment) as discrete chemicals from a toxicological perspective. The aggregate effect of these metabolites and degradation products is usually considered to be no worse than the effect from the original amount of parent material. However, in some instances these substances may pose a greater risk to the environment than the associated parent compound (Boxall et al. 2004a).

Information on major metabolites is typically collected as part of efficacy or safety studies. Furthermore, there are increasingly robust computational approaches for predicting likely metabolite and degradate profiles from parent structures (e.g., Mekenyan et al. 2005). Given the availability of empirical data for metabolites or models that can predict both metabolites and degradates, it seems reasonable that this information be used in some fashion to assess possible aquatic hazards. However, it clearly is not feasible from a resource perspective to conduct separate aquatic toxicity studies for all observed and predicted metabolites or degradates. If there were an approach to identify a subset of derivative structures with the potential to cause unacceptable toxicity, additional fate and effects testing with these chemicals could be a reasonable option. Simple rule-based approaches have been proposed for assessing pesticide degradates (e.g., Sinclair and Boxall 2003). Moreover, a system is available that features a simulator of environmental degradates linked to quantitative-structure activity relationship (QSAR) models designed for different toxicity pathways. The system, which predicts specific degradates or metabolites, is linked to several aquatic QSARs allowing, for example, the prediction of baseline toxicity (to fish) for each metabolite based on a narcosis MOA (Veith et al. 1983) or the occurrence of degradates with possible reactive MOA (e.g., electrophiles) that would result in greater than baseline toxicity. Additionally, USEPA researchers and collaborators are developing a metabolism simulator to identify chemicals that are metabolically activated (in vertebrates) to forms that can bind to specific physiological receptors using receptor-binding QSARs (Kolanczyk et al. 2005; Mekenyan et al. 2005; Serafimova et al. 2005;

Todorov et al. 2005). Outputs of the linked models would not, by themselves, be suitable for making predictions of metabolite or degradate toxicity, but the information could help identify where further testing could be warranted. Although not currently widely distributed, versions of this linked metabolism and toxicity model should be available soon.

ADME studies may also provide valuable information, as in these studies parent and metabolites occur at approximately the same time or sequentially in the plasma of treated animals. Although the in vitro activity of primary metabolites for the target effect might be known, the contribution of the metabolites to the toxicological response in mammalian studies is not normally investigated. However, when the plasma concentrations of metabolites are known at doses resulting in no effects in mammals, it could be informative to compare those plasma concentrations to calculated plasma concentrations in aquatic vertebrates potentially exposed to the compounds in surface waters. The known concentration of metabolites in mammals does not currently provide a reasonable starting point to extrapolate the effects of metabolites to aquatic organisms, but it would be useful to understand concentrations at which no effects were found. In the future, it may be possible to relate information from target animal ADME profiles or mammalian toxicology ADME profiles to identify the effects of major metabolites and use that information as a starting point for identifying effects from metabolites in aquatic vertebrates using structure-activity relationship analysis.

5.5.4 MIXTURES

It is highly unlikely that veterinary medicines will exist in aquatic systems alone, so it is important that the potential interactions with other veterinary medicines and contaminants are considered. Aquatic hazards of human medicinal mixtures have been reviewed by Mihaich et al. (2005). Evaluation of the hazard from exposure to chemical mixtures is complex, in part due to the potential combinations and concentrations of chemicals that could already be in surface water. These chemicals can have a variety of toxic mechanisms. Prospective evaluations of an additional chemical moving into surface water that already has a large number of possible combinations and trace concentrations of natural and anthropogenic molecules are too numerous to conduct, even if test data were available. Fortunately, whole effluent testing and conceptually similar evaluation techniques exist for waste mixtures that could be of special concern. But the toxicity of even these mixtures is generally driven by the potency and concentration of 1 or 2 chemicals.

There may be special circumstances where it is interesting to try to evaluate the hazard of simple mixtures, but even this is not straightforward. Normally, when mechanisms of toxicity are different or unknown for 2 chemicals it is reasonable to assume that the result of being exposed to both is equivalent to being exposed independently to each one. The chemicals in the simple mixture must have the same mode of action, potentially even competing for the same receptor or molecular target, in order to project realistically the potential for an increased target organism response to exposure to the mixture. Even with the same mode of action, the most likely description of the additive hazards from 2 chemicals would only be concentration addition (Williams 2005). The application factors (at least 10 ×) used on the most sensitive chronic NOEC test values to predict a chronic no-effect concentration for all aquatic organisms for each individual chemical may be adequate to account for combined increases in hazard.

Estimation of the mixture hazards to aquatic organisms from veterinary medicine products that are combinations of 2 active ingredients should normally be performed as if the organisms were exposed to each compound independently. The active ingredients would normally not be placed in 1 product if they competed for the same molecular target, so the probability of increased aquatic organism sensitivity due to a mixture effect is low. The exception to this could be situations where 2 drugs act in a nonadditive manner (e.g., a potentiated sulfonamide) or a retrospective risk assessment of a CAFO-impacted watershed, if exposure modeling predicts and field assessment confirms co-occurrence of multiple compounds with different modes of action, particularly in watersheds with high densities of CAFOs. In these scenarios it may be useful to screen potential mixture interactions of the co-occurring chemicals as either 1) compounds with a similar MOA, which may result in additive responses or 2) compounds with drug–drug interaction profiles in mammals or targeted livestock, which may result in nonadditive responses.

Toxicogenomic data also might be useful for dealing with situations in which organisms are expected to be exposed to mixtures of veterinary medicines that may, or may not, have similar MOAs. There is a solid toxicological basis (and regulatory precedence) for using concentration addition models to predict the joint toxicity of mixtures of chemicals with a common MOA. Molecular profiling or fingerprinting data can logically be used as a basis for "binning" chemicals with similar MOAs to decide when concentration addition is a viable approach for dealing with a veterinary medicine mixture. This type of approach should become increasingly viable as toxicogenomic data are collected and archived from base tests with a variety of veterinary medicines. As further discussed in Section 5.5.7, lotic and lentic mesocosms are useful for higher tiered assessment of contaminant mixture hazards in aquatic ecosystems.

5.5.5 Enantiomer-Specific Hazard

In recent years, increased attention has been given to chiral molecules in the environment (Garrison 2006). A number of environmental contaminants including representatives from human and veterinary medicine and pesticide classes are chiral compounds that are distributed as racemic mixtures; a racemic mixture is a 1:1 mixture of enantiomers. For example, all synthetic pyrethroid insecticides are chiral. Although enantiomers have identical physiochemical properties and molecular formulae (Kallenborn and Hühnerfuss 2001), they may differ in environmental fate, bioavailability, potency, and toxicity due to stereospecific biological receptors (Mathison et al. 1989; Ali and Aboul-Enein 2004).

Whereas information for enantiomer-specific fate and effects is limited in the literature, the majority of peer-reviewed information on enantiospecific environmental fate has been characterized for pesticides (Hegeman and Laane 2002). Recent studies by Fono and Sedlak (2004) and Nikolai et al. (2006) examining beta-adrenergic receptor blocker medicines in municipal wastewater found enantiomer-specific degradation such that the ratio of enantiomers deviated from the 1:1 racemic mixture. In addition to such enantiospecific degradation, which can influence ambient exposure to chiral veterinary medicines, internal dose of chiral contaminants may be influenced by enantiospecific differences in metabolism and clearance rates. For example, increased clearance of 1 enantiomer of a racemic veterinary medicine may lead to enantiomerspecific differences in accumulation in the tissues of exposed organisms (Hummert et al. 1995).

A limited number of studies have examined aquatic toxicity of enantiomers of select pyrethroid and organophosphate insecticides to the cladocerans *Daphnia magna* and *Ceriodaphnia dubia* (Yen et al. 2003; Wang et al. 2004; Liu et al. 2005a, 2005b). Although these investigations evaluated only acute mortality responses, differences in LC50 values between enantiomers ranged from 3 to ~40 fold, indicating substantial enantiospecific toxicity by representative compounds from classes of veterinary medicines. More recently, Stanley et al. (2006) extended these findings with pesticides to medicines by demonstrating that the most potent enantiomer of propranolol in mammals was most sublethally toxic to the model fish *Pimephales promelas*, but not to *D. magna*, potentially because cladocerans do not have pharmacological targets (e.g., beta-adrenergic receptors) for propranolol.

Enantiospecific differences in fate and effects for chiral contaminants are often ignored in exposure and effect analyses of ecological risk assessments. It is most common to treat a mixture of enantiomers as 1 compound in prospective and retrospective assessments of chiral molecules, largely because of limited published studies on enantiospecific fate and effects. However, if enantiospecific differences in fate and effects occur between enantiomers, consistent with those described above, uncertainty may be unnecessarily introduced into ecological risk assessments of chiral veterinary medicines.

5.5.6 SORPTION TO SEDIMENT

There is increasing evidence from monitoring studies that some classes of veterinary medicines can concentrate in aquatic sediments (see Chapter 4). This clearly is an important observation in terms of ascertaining the overall fate and transport of veterinary medicines in the environment. However, occurrence of veterinary medicines in sediments also has potential repercussions as to how best to assess their hazard and effects.

If there is an indication that a veterinary medicine could accumulate in sediment, it is logical to be concerned about possible effects in benthic species, typically invertebrates. In prospective assessments it is possible to assess potential effects of sediment-associated contaminants using either an empirical or predictive approach. The empirical approach would feature spiking test sediment(s) with different concentrations of the veterinary medicine of interest followed by short- or long-term toxicity tests with freshwater or marine benthic invertebrates (e.g., amphipods, chironomids, oligochaetes, or polychaetes). There are standard methods available for conducting these types of toxicity tests, as well as information on spiking sediments (US Environmental Protection Agency [USEPA] 1994a, 1994b, 2001; Organization for Economic Cooperation and Development [OECD] 218, 2004a; OECD 219, 2004b). Although the empirical approach certainly will yield data, it can be technically very challenging (and expensive) in terms of sediment spiking and exposure characterization. Furthermore, many scientists and risk assessors consider the results of spiked-sediment tests limited in terms of extrapolation to sediments with characteristics (e.g., organic carbon content) different from those actually tested. Hence, a predictive approach to assessing potential effects of sediment-associated veterinary medicines to benthic invertebrates might be preferable, at least as a complement to sediment toxicity tests. One approach may be to use equilibrium partitioning theory (EqP) (Di Toro et al. 1991; USEPA 1993). This approach can utilize water-only toxicity data (such as those collected for the base veterinary medicine tests), in conjunction with the EqP model, to predict which chemical and sediment combinations would present unacceptable risks to aquatic invertebrates. This approach assumes that partitioning between water and sediment is governed by hydrophobicity and the organic carbon content of the sediment. However, there is an increasing body of evidence demonstrating that this assumption is invalid for many veterinary medicine classes (Tolls 2001); for example, enantiospecific (Wedyan and Preston 2005) and pH-dependent partitioning for ionizable compounds (Kummerer et al. 2005) can influence sorption to sediments. In these cases, other modeling approaches should be explored.

An additional consideration from the standpoint of sediment-associated veterinary medicines involves bioaccumulation by benthic animals and subsequent food chain transfer, which could result in secondary (or transfer) toxicity in organisms (fish, birds, or mammals) consuming aquatic invertebrates. Again it is possible to approach this from either an empirical or predictive perspective. There are tests available for both saltwater and freshwater species to measure bioaccumulation directly. For example, USEPA (1994a) describes a method to determine bioaccumulation of chemicals using freshwater oligochaetes exposed to either spiked or field-collected sediments. However, as is the case for toxicity tests, there are some significant technical and resource challenges associated with the spiked-sediment studies. Modeling approaches would be a valuable complement.

5.5.7 Assessing Effects on Communities

Many of the studies currently recommended for veterinary medicines involve tests on single organisms. They therefore ignore the complex interactions (including indirect effects) that can occur in the real environment. Multispecies responses and indirect effects that are mediated through species interactions can be

TABLE 5.6
Typical types and characteristics of cosms

Type	Size	Number of trophic levels	Length of time used	Location
Nanocosm	1–100 L	2	< 8 weeks	Indoor
Microcosm	100–15,000 L	≥ 3	1 season	Indoor or outdoor
Mesocosm	> 15,000 L	≥ 3	> 1 season	Outdoor

addressed in studies conducted in artificial multispecies test systems (microcosms or mesocosms) or in the field. The suffix "-cosm" generally refers to a wide variety of experimental systems (Table 5.6), ranging from small laboratory flasks to large outdoor streams, tanks, or ponds. The distinguishing feature of cosms is the inclusion of multiple ecological components (species, functional groups, or habitat types) to simulate ecological processes as they occur in nature. Cosms bridge the gap between simple laboratory test systems and full-scale field studies and can be used to test hypotheses suggested by observations in laboratory studies or from other knowledge. Cosm studies provide effect measures that can be closer to the assessment measures, for the following reasons (Solomon et al. 1996):

- Measurements of productivity in cosms incorporate the aggregate responses of multiple species — often several dozen — in each trophic level. Because organisms can vary widely in their sensitivity to the stressor, the overall response of the community may be quite different from the responses of individual species as measured in laboratory toxicity tests.
- Cosm studies allow observation of population and community recovery from the effects of the stressor.
- Studies with cosms allow measurement of indirect effects of stressors on other trophic levels. Indirect effects may result from changes in food supply, habitat, or water quality. Such effects may be inferred by extrapolation from laboratory toxicity data, but they can be measured directly only in multitrophic systems.
- Cosm studies can be designed to approximate realistic stressor exposure regimes more closely than standard laboratory single-species toxicity tests. Most studies, especially those conducted in outdoor systems, incorporate partitioning, degradation, and dissipation — important factors in determining exposure. These factors are rarely accounted for in laboratory toxicity studies but may greatly influence the magnitude of ecological response.

A number of procedures have been proposed for cosm types of test (Arnold et al. 1991), and there are numerous examples of their utility (Hill et al. 1994). Most of this work has been carried out in aquatic systems on a range of substances

(Giesy 1980; Giddings 1983; Franco et al. 1984; Solomon et al. 1996; Giesy et al. 1999; Giddings et al. 2000, 2001, 2005), including human and veterinary medicines (Brain et al. 2004, 2005a, 2005b; Richards et al. 2004; Wilson et al. 2004; Van den Brink et al. 2005).

A number of cosm studies have been performed on active ingredients used as veterinary medicines. Indirect effects mediated through the food chain have been observed for pyrethroids, some of which are used in veterinary applications (Kaushik et al. 1985; Giddings et al. 2001). In this case, insensitive rotifers increased in numbers when populations of more sensitive Cladocera and other crustaceans declined. Another indirect response observed in a cosm is that of photosynthesis inhibition through light adsorption by degradates of tetracyclines (Brain et al. 2005b). This response was mediated via a physical process and was observed only at high concentrations and in the absence of hydraulic dilution such as may be present in lotic systems.

Effect classes that can be used to summarize observed effects in aquatic cosm studies are described in the EU Guidance Document on Aquatic Ecotoxicology (Brock et al. 2000a, 2000b; Health and Consumer Affairs [SANCO] 2002). In Europe, these effect classes are used to evaluate semifield tests submitted for the registration of pesticides, but they could also be adapted to the assessment of veterinary medicines, as discussed in Chapter 3.

Although the potential adverse effects of veterinary medicines may be observed in stream ecosystems, no studies have assessed their effects on structural or functional response variables in model stream systems. Lotic systems involve physical, chemical, and biotic characteristics that vary greatly from lentic systems; variation in these characteristics could influence the exposure and effects of veterinary medicines (e.g., leaf litter breakdown by detritivores and microorganisms). Thus, it is likely that a robust understanding of cause-and-effect relationships between environmentally realistic veterinary medicine exposures and stream ecosystem functional responses will only be possible if sophisticated lotic mesocosms are employed (Brooks et al. 2007). Furthermore, it is likely that many cases of veterinary medicine contamination in streams could be accompanied by significant nutrients and particulate matter loads due to co-contamination by either sewage or animal waste. Here again, appropriately designed stream mesocosm experiments coupled with other lines of evidence can be used to characterize chemical stressor effects on aquatic ecosystems appropriately (Brooks et al. 2004; Stanley et al. 2005).

5.6 CONCLUSIONS

There is increasing interest in the potential impacts of veterinary medicines on aquatic ecosystems. In recent years significant progress has been made in the development and standardization of ecological risk assessment methodologies for these substances. These methodologies tend to involve the use of standard test organisms (fish, daphnids, and algae) and endpoints (mortality, growth, or reproduction) in the laboratory. However, veterinary medicines are biologically active substances; there is an increasing body of evidence that exposure to select medicine groups could result in effects not identified using standard methodologies, and it is possible that indirect effects could be elicited. Moreover, the exposure profiles and bioavailability of veterinary medicines in lentic and lotic aquatic systems will be very different than in the laboratory. By combining information on a substance's efficacy and toxicology in target organisms and mammals with ecotoxicogenomic approaches and higher tier assessment approaches developed for pesticides, it should be possible to develop a more complete understanding of the real risks of veterinary medicines to aquatic systems. Many of these approaches also have potential applications in retrospective assessment work such as postauthorization monitoring, watershed assessments, and toxicity identification evaluations.

Our current understanding of some issues is poorly developed, and we would recommend that future efforts focus on a number of key areas, namely the following:

1) Accumulation of data and knowledge to test and further refine extrapolations from mammalian toxicity data to aquatic effects
2) Further development of ecotoxicogenomic approaches and exploration of how data from these efforts can be applied in prospective and retrospective risk assessment
3) Development and validation of methods for metabolite and degradate assessment
4) Studies to understand further those factors and processes affecting the bioavailability and trophic transfer of veterinary medicines in aquatic systems
5) Studies to examine influences of enantiomer-specific fate and effects of racemic veterinary medicines
6) Consideration of the potential impacts of mixtures of veterinary medicines and mixtures containing veterinary medicines and other contaminant classes (e.g., nutrients)

It is also important that regulatory guidance documents be reviewed on a regular basis to account for developments in these and other areas.

REFERENCES

Ali I, Aboul-Enein HY. 2004. Chiral pollutants: distribution, toxicity, and analysis by chromatography and capillary electrophoresis. Chichester (UK): John Wiley.

Ankley GT, Black MC, Garric J, Hutchinson TH, Iguchi T. 2005. A framework for assessing the hazard of pharmaceutical materials to aquatic species. In: Williams R, editor. Science for assessing the impacts of human pharmaceutical materials on aquatic ecosystems. Pensacola (FL): SETAC Press, p 183–238.

Ankley GT, Daston G, Degitz S, Denslow N, Hoke R, Kennedy S, Miracle A, Perkins E, Snape J, Tillitt D, Tyler C, Versteeg D. 2006. Toxicogenomics in regulatory ecotoxicology: potential applications and practical challenges. Environ Sci Technol 40:4055–4065.

Ankley GT, Jensen KM, Kahl MD, Korte JJ, Makynen EA. 2001. Description and evaluation of a short-term reproduction test with the fathead minnow (Pimephales promelas). Environ Toxicol Chem 20:1276–1290.

Ankley GT, Jensen KM, Makynen EA, Kahl MD, Korte JJ, Hornung MW, Henry TR, Denny JS, Leino RL, Wilson VS, Cardon MC, Hartig PC, Gray LE. 2003. Effect of the androgenic growth promoter 17-β-Tenbolone on fecundity and reproductive endocrinology of the fathead minnow. Environ Tox Chem 22:1350–1360.

Arnold D, Hill IR, Matthiessen P, Stephenson R, editors. 1991. Guidance document on testing procedures for pesticides in freshwater mesocosms. Brussels (Belgium): SETAC-Europe.

Boxall ABA, Blackwell PA, Cavallo R, Kay P, Tolls J. 2002a. The sorption and transport of a sulphonamide antibiotic in soil systems. Toxicol Lett 131:19–28.

Boxall ABA, Brown CD, Barrett K. 2002b. Higher tier aquatic toxicity testing for pesticides. Pest Man Sci 58:637–648.

Boxall ABA, Fenner K, Kolpin DW, Maund S. 2004a. Environmental degradates of synthetic chemicals: fate, effects and potential risks. Environ Sci Technol 38:369A–375A.

Boxall ABA, Fogg LA, Blackwell PA, Kay P, Pemberton EJ, Croxford A. 2004b. Veterinary medicines in the environment. Rev Environ Contam Toxicol 180:1–91.

Boxall ABA, Kolpin D, Halling-Sørensen B, Tolls J. 2003. Are veterinary medicines causing environmental risks? Environ Sci Technol 37:286A–294A.

Brain RA, Bestari KT, Sanderson H, Hanson ML, Wilson CJ, Johnson DJ, Sibley PK, Solomon KR. 2005a. Aquatic microcosm assessment of the effects of tylosin to Lemna gibba and Myriophyllum spicatum. Environ Pollut 133:389–401.

Brain RA, Johnson DJ, Richards RA, Hanson ML, Sanderson H, Lam MW, Mabury SA, Sibley PK, Solomon KR. 2004. Ecotoxicological evaluation of the effects of an eight pharmaceutical mixture to the aquatic macrophytes Lemna gibba and Myriophyllum sibiricum exposed in aquatic microcosms. Aquat Toxicol 70:23–40.

Brain RA, Sanderson H, Sibley PK, Solomon KR. 2006. Probabilistic ecological hazard assessment: evaluation pharmaceutical effects on aquatic higher plants as an example. Ecotox Environ Saf 64:128–135.

Brain RA, Wilson CJ, Johnson DJ, Sanderson H, Bestari K, Hanson ML, Sibley PK, Solomon KR. 2005b. Effects of a mixture of tetracyclines to Lemna gibba and Myriophyllum sibiricum evaluated in aquatic microcosms. Environ Pollut 138:426–443.

Brock TCM, Lahr J, Van den Brink PJ. 2000a. Ecological risks of pesticides in freshwater ecosystems part 1: herbicides. Wageningen (The Netherlands): Alterra, 124 p

Brock TCM, Van Wijngaarden RPA, Van Geest GJ. 2000b. Ecological risks of pesticides in freshwater ecosystems part 2: insecticides. Wageningen (The Netherlands): Alterra, 141 p.

Brooks BW, Foran CM, Richards SM, Weston J, Turner PK, Stanley JK, Solomon KR, Slattery M, La Point TW. 2003. Aquatic ecotoxicology of fluoxetine. Toxicol Lett 142:169–183.

Brooks BW, Maul J, Belden J. 2007. Emerging contaminants: antibiotics in aquatic and terrestrial ecosystems. In: Jorgensen SE, editor. Encyclopedia of ecology. London (UK): Elsevier Press.

Brooks BW, Riley T, Taylor RD. 2006. Water quality of effluent-dominated stream ecosystems: ecotoxicological, hydrological, and management considerations. Hydrobiologia 556:365–379.

Brooks BW, Stanley JK, White JC, Turner PK, Wu KB, La Point TW. 2004. Laboratory and field responses to cadmium in effluent-dominated stream mesocosms. Environ Toxicol Chem 24:464–469.

Cairns J Jr. 1983. Are single species toxicity tests alone adequate for estimating environmental hazard? Hydrobiologia 100:47–57.

Cairns J Jr. 1986. The myth of the most sensitive species. BioScience 36:670–672.

Cassarett J, Amdur MO, Klaassen C. 1986. Cassarett and Doull's toxicology: the basic science of poisons. 2nd ed. New York (NY): Macmillan.

Crane M, Watts CW, Boucard T. 2006. Chronic aquatic environmental risks from exposure to human pharmaceuticals. Sci Total Environ 367:23–41.

Daughton CG, Ternes TA. 1999. Pharmaceuticals and personal care products in the environment: agents of subtle change? Special report. Environ Health Perspect Suppl 107:907–938.

de Schamphelaere KAC, Janssen CR. 2002. A biotic ligand model predicting acute copper toxicity to Daphnia magna: the effects of calcium, magnesium, sodium, potassium and pH. Environ Sci Technol 36:48–54.

Dickson KL, Waller WT, Kennedy JH, Ammann LP. 1992. Assessing the relationship between ambient toxicity and instream biological response. Environ Toxicol Chem 11:1307–1322.

Di Toro DM, Zarba CS, Hansen DJ, Berry WJ, Swartz RC, Cowan CE, Pavlou SP, Allen HE, Thomas NA, Paquin, PR. 1991. Technical basis for establishing sediment quality criteria for nonionic organic chemicals using equilibrium partitioning. Environ Toxicol Chem 10:5141–5183.

Fono L, Sedlak DL. 2004. Use of the chiral pharmaceutical propranolol to identify sewage discharges into surface waters. Environ Sci Technol 39:9244–9252.

Franco PJ, Giddings JM, Herbes SE, Hook LA, Newbold JD, Roy WK, Southworth GR, Stewart AJ. 1984. Effects of chronic exposure to coal-derived oil on freshwater ecosystems. I. Microcosms. Environ Toxicol Chem 3:447–463.

Garrison AW. 2006. Probing the enantioselectivity of chiral pesticides. Environ Sci Technol 40:16–23.

Giddings JM. 1983. Microcosms for assessment of chemical effects on the properties of aquatic ecosystems. In: Saxena J, editor. Hazard assessment of chemicals — current developments. Vol. 2. New York (NY): Academic Press, p 45–94.

Giddings JM, Anderson TA, Hall LW, Jr, Kendall RJ, Richards RP, Solomon KR, Williams WM. 2005. A probabilistic aquatic ecological risk assessment of atrazine in North American surface waters. Pensacola (FL): SETAC Press, 432 p.

Giddings JM, Hall LWJ, Solomon KR. 2000. An ecological risk assessment of diazinon from agricultural use in the Sacramento–San Joaquin River basins, California. Risk Anal 20:545–572.

Giddings JM, Solomon KR, Maund SJ. 2001. Probabilistic risk assessment of cotton pyrethroids: II. Aquatic mesocosm and field studies. Environ Toxicol Chem 20:660–668.

Giesy JP. 1980. Microcosms in ecological research. CONF–781101. Washington (DC): US Department of Energy.

Giesy JP, Solomon KR, Coates JR, Dixon KR, Giddings JM, Kenaga EE. 1999. Chlorpyrifos: ecological risk assessment in North American aquatic environments. Rev Environ Contam Toxicol 160:1–129.

Hall LWJ, Anderson RD. 1995. The influence of salinity on the toxicity of various classes of chemicals to aquatic biota. Crit Rev Toxicol 25:281–346.

Halling-Sørensen B. 1999. Algal toxicity of antibacterial agents used in intensive farming. Chemosphere 40:731–739.

Health and Consumer Affairs [SANCO]. 2002. Guidance document on aquatic eco-toxicology in the context of directive 91/414/EEC. SANCO Report 3268_rev 3. http://ec.europa.eu/food/plant/protection/evaluation/guidance/wrkdoc10_en.pdf.

Hegeman WJM, Laane RWPM. 2002. Enantiomeric enrichment of chiral pesticides in the environment. Rev Environ Contam Toxicol 173:8–116.

Hill IR, Heimbach F, Leeuwangh P, Matthiessen P, editors. 1994. Freshwater field tests for hazard assessment of chemicals. Boca Raton (FL): CRC Press, 561 p.

Hobson J, Endris R, Wislocki P. 2004. Ecological risk assessment of the sea lice control agent emamectin benzoate. 25th annual North American meeting of the Society of Environmental Toxicology and Chemistry, Portland, Oregon, November 14–18.

Hoke RA, Ankley GT, Cotter AM, Goldenstein T, Kosian PA, Phipps GL, VanderMeiden FM. 1994. Evaluation of equilibrium partitioning theory for predicting acute toxicity of field-collected sediments contaminated with DDT, DDE, and DDD to the amphipod Hyalella azteca. Environ Toxicol Chem 13:157–166.

Huggett DB, Brooks BW, Peterson B, Foran CM, Schlenk D. 2002. Toxicity of select beta-adrenergic receptor blocking pharmacueticals (β-blockers) on aquatic organisms. Arch Environ Contam Toxicol 42:229–235.

Huggett DB, Cook JC, Ericson JF, Williams RT. 2003. A theoretical model for utilizing mammalian pharmacology and safety data to prioritized potential impacts of human pharmaceuticals to fish. Hum Ecol Risk Ass 9:1789–1800.

Hummert K, Vetter W, Luckas B. 1995. Levels of alpha-HCH, and enantiomeric ratios of alpha-HCH in marine mammals from the northern hemisphere. Chemosphere 31:3489–3500.

Ingerslev F, Halling-Sørensen B. 2001. Biodegradability of metronidazole, olaquindox and tylosin and formation of tylosin degradation products in aerobic soil/manure slurries. Chemosphere 48:311–320.

Jørgensen SE, Halling-Sørensen B. 2000. Drugs in the environment. Chemosphere 40:691–699.

Kallenborn R, Hühnerfuss H. 2001. Chiral environmental pollutants, trace analysis and ecotoxicology. Berlin (Germany): Springer-Verlag.

Kaushik NK, Stephenson GL, Solomon KR, Day KE. 1985. Impact of permethrin on zooplankton communities using limnocorrals. Can J Fish Aquat Sci 42:77–85.

Kay P, Blackwell PA, Boxall ABA. 2004. Fate and transport of veterinary antibiotics in drained clay soils. Environ Toxicol Chem 23:1136–1144.

Kolanczyk R, Tapper M, Nelson B, Wehninger V, Denny J, Kuehl D, Sheedy B, Mazur C, Kenneke J, Jones J, Schmieder P. 2005. Increased endocrine activity of xenobiotic chemicals as mediated by metabolic activation. 26th annual meeting of the Society of Environmental Toxicology and Chemistry, Baltimore, Maryland, November 13–17.

Koschorreck J, Koch C, Ronnefahrt I. 2002. Environmental risk assessment of veterinary medical products in the EU: a regulatory perspective. Toxicol Lett 131:117–124.

Kümmerer K, Ericson J, Hannah B, Johnson A, Sedlak DL, Weston JJ. 2005. Environmental fate and transport of human pharmaceuticals. In: Williams R, editor. Science for assessing the impacts of human pharmaceutical materials on aquatic ecosystems. Pensacola (FL): SETAC Press, p 111–148.

La Point TW, Waller WT. 2000. Field assessment in conjunction with whole effluent toxicity testing. Environ Toxicol Chem 19:14–24.

Lettieri T. 2006. Recent applications of DNA microarray technology to toxicology and ecotoxicology. Environ Health Perspect 114:4–9.

Leung KMY, Morritt D, Wheeler JR, Whitehouse P, Sorokin N, Toy R, Holt R, Crane M. 2002. Can saltwater toxicity be predicted from freshwater data? Mar Pollut Bull 42:1007–1013.

Liu W, Gan J, Lee S, Werner I. 2005a. Isomer selectivity in aquatic toxicity and biodegradation of bifenthrin and permethrin. Environ Toxicol Chem 24:1861–1866.

Liu W, Gan J, Schlenk D, Jury WA. 2005b. Enantioselectivity in environmental safety of current chiral insecticides. Proceedings of the National Academy of Sciences of the United States of America 102:701–706.

MacGregor JT. 2003. The future of regulatory toxicology: impact of the biotechnology revolution. Toxicol Sci 75:236–248.

Mathison IW, Solomons WE, Morgan PH, Tidwell RR. 1989. Structural features and pharmacologic activity. In: Foye WO, editor. Principles of medicinal chemistry. Philadelphia (PA): Lea & Febiger, p 49–77.

Mekenyan O, Jones J, Schmieder P, Kotov S, Pavlov T, Dimitrov S. 2005. Performance, reliability and improvement of a tissue-specific metabolic simulator. 26th annual meeting of the Society of Environmental Toxicology and Chemistry, Baltimore, Maryland, November 13–17.

Mihaich EM, Borgert CJ, Brighty GC, Kortenkamp A, Laenge R, Snyder SA, Sumpter JP. 2005. Evaluating simple and complex mixtures containing pharmaceuticals in the environment. In: Williams R, editor. Science for assessing the impacts of human pharmaceutical materials on aquatic ecosystems. Pensacola (FL): SETAC Press, p 239–268.

Nikolai LN, McClure EL, MacLeod SL, Wong CS. 2006. Stereoisomer quantification of the β-blocker drugs atenolol, metoprolol, and propranolol in wastewaters by chiral high-performance liquid chromatography–tandem mass spectrometry. J Chromatogr A 1131:103–109.

Nuwaysir EF, Bittner M, Trent J, Barrett JC, Afshari CA. 1999. Microarrays and toxicology: the advent of toxicogenomics. Mol Carcinog 24:153–159.

[OECD] Organization for Economic Cooperation and Development. 2004a. Sediment-water chironomid toxicity test using spiked sediment. OECD 218. Paris (France): OECD, 21 p.

[OECD] Organization for Economic Cooperation and Development. 2004b. Sediment-water chironomid toxicity test using spiked water. OECD 219. Paris (France): OECD, 21 p.

Richards SM, Wilson CJ, Johnson DJ, Castle DM, Lam M, Mabury SA, Sibley PK, Solomon KR. 2004. Effects of pharmaceutical mixtures in aquatic microcosms. Environ Toxicol Chem 23:1035–1042.

Santore RC, Di Toro DM, Paquin PR, Allen HE, Meyer JS. 2001. Biotic ligand model of the acute toxicity of metals. 2. Application to acute copper toxicity in freshwater fish and Daphnia. Environ Toxicol Chem 20:2397–2402.

Schmidt CW. 2004. Metabolomics: what's happening downstream of DNA. Environ Health Perspect 112:A410–A415.

Serafimova R, Aladjov R, Kolanczyk R, Schmieder P, Akahori Y, Nakai M, Jones J, Mekenyan O. 2005. QSAR evaluation of ER binding affinity of chemicals and metabolites. 26th annual meeting of the Society of Environmental Toxicology and Chemistry, Baltimore, Maryland, November 13–17.

[SETAC] Society of Environmental Toxicology and Chemistry. 1994. Final report: Aquatic Risk Assessment Mitigation Dialogue Group. Pensacola (FL): SETAC Press, p 26–63.

Sinclair CJ, Boxall ABA. 2003. Assessing the ecotoxicity of pesticide transformation products. Environ Sci Technol 37:4617–4625.

Solomon KR, Baker DB, Richards P, Dixon KR, Klaine SJ, La Point TW, Kendall RJ, Giddings JM, Giesy JP, Hall LWJ, Weisskopf C, Williams M. 1996. Ecological risk assessment of atrazine in North American surface waters. Environ Toxicol Chem 15:31–76.

Solomon K, Takacs P. 2002. Probability risk assessment using species sensitivity distributions. In: Postluma L, Suter G, Traas T, editors. Species sensitivity distributions in ecotoxicology. Boca Raton (FL): CRC Press, p 286–313.

Stanley JK, Brooks BW, La Point TW. 2005. A comparison of chronic cadmium effects on Hyalella azteca in effluent-dominated stream mesocosms to similar laboratory exposures in effluent and reconstituted hard water. Environ Toxicol Chem 24:902-908.

Stanley JK, Ramirez AJ, Mottaleb M, Chambliss CK, Brooks BW. 2006. Enantiospecific toxicity of the β-blocker propranolol to Daphnia magna and Pimephales promelas. Environ Toxicol Chem 25:1780–1786.

Suter G. 2002. North American history of species sensitivity distributions. In: Posthuma L, Suter G, Traas T, editors. Species sensitivity distributions in ecotoxicology. Boca Raton (FL): CRC Press, p 11–17.

Telfer T, Baird D, McHenery J, Stone J, Sutherland I, Wislocki P. 2006. Environmental effects of the anti-sea lice (Copepoda: Caligidae) therapeutant emamectin benzoate under commercial use conditions in the marine environment. Aquaculture 260:163–180.

Todorov M, Serafimova R, Schmieder P, Aladjov H, Mekenyan O. 2005. A QSAR evaluation of AR binding affinity of chemicals. 26th annual meeting of the Society of Environmental Toxicology and Chemistry, Baltimore, Maryland, November 13–17.

Tolls J. 2001. Sorption of veterinary medicines: a review. Environ Sci Technol 35:3397–3406.

[USEPA] US Environmental Protection Agency. 1993. Technical basis for deriving sediment quality criteria for nonionic organic chemicals for the protection of benthic organisms using equilibrium partitioning. EPA/822/R-93/011. Washington (DC): US Environmental Protection Agency.

[USEPA] US Environmental Protection Agency. 1994a. Methods for measuring the toxicity and bioaccumulation of sediment-associated contaminants with freshwater invertebrates. EPA/600/R-94/024. Duluth (MN): US Environmental Protection Agency.

[USEPA] US Environmental Protection Agency. 1994b. Methods for measuring the toxicity of sediment-associated contaminants with estuarine and marine amphipods. EPA/600/R-94/025. Narragansett (RI): US Environmental Protection Agency.

[USEPA] US Environmental Protection Agency. 2001. Methods for collection, storage and manipulation of sediment for chemical and toxicological analyses: technical manual. EPA/823/B-01/002. Washington (DC): US Environmental Protection Agency.

Van den Brink PJ, Tarazona JV, Solomon KR, Knacker T, Van den Brink NW, Brock TCM, Hoogland JPH. 2005. The use of terrestrial and aquatic microcosms and mesocosms for the ecological risk assessment of veterinary medicinal products. Environ Toxicol Chem 24:820–829.

Veith GD, Call DJ, Brooke LT. 1983. Structure-toxicity relationships for the fathead minnow, Pimephales promelas: narcotic industrial chemicals. J Fish Res Bd Can 40:743–748.

[VICH] International Cooperation on Harmonization of Technical Requirements for Registration of Veterinary Products [VICH]. 2000. Environmental impact assessment (EIAs) for veterinary medicinal products (veterinary medicines) — phase 1. VICH GL6 (ecotoxicity phase 1), June 2000, for implementation at phase 7. Brussels (Belgium): VICH.

[VICH] International Cooperation on Harmonization of Technical Requirements for Registration of Veterinary Products [VICH]. 2004. Environmental impact assessment for veterinary medicinal products phase II guidance. VICH-GL38. Brussels (Belgium): VICH, 38 p. http://vich.eudra.org.

Wang YS, Tai KT, Yen JH. 2004. Separation, bioactivity, and dissipation of enantiomers of the organophosphorus insecticide fenamiphos. Ecotox Environ Saf 57:346–353.

Waters MD, Fostel JM. 2004. Toxicogenomics and systems toxicology: aims and prospects. Nature Rev Genetics 5:936–948.

Wedyan M, Preston MR. 2005. Isomer-selective adsorption of amino acids by components of natural sediment. Environ Sci Technol 39:2115–2119.

Wheeler JR, Grist EPM, Leung KMY, Morritt D, Crane M. 2002. Species sensitivity distributions: data and model choice. Mar Pollut Bull 45:192–202.

Williams RT. 2005. Human health pharmaceuticals in the environment — an introduction. In: Williams RT, editor. Human pharmaceuticals: assessing the impacts on aquatic systems. Pensacola (FL): SETAC Press, p. 1–46.

Wilson CJ, Brain RA, Sanderson H, Johnson DJ, Bestari KT, Sibley PK, Solomon KR. 2004. Structural and functional responses of plankton to a mixture of four tetracyclines in aquatic microcosms. Environ Sci Technol 38:6430–6439.

Yen JH, Tsai CC, Wang YS. 2003. Separation and toxicity of enantiomers of organophosphorus insecticide leptophos. Ecotox Environ Saf 55:236–242.

6 Exposure Assessment of Veterinary Medicines in Terrestrial Systems

Louise Pope, Alistair Boxall, Christian Corsing,
Bent Halling-Sørensen, Alex Tait, and
Edward Topp

6.1 INTRODUCTION

It is inevitable that during their use, veterinary medicines will be released to the terrestrial environment. For hormones, antibiotics, and other pharmaceutical agents administered either orally or by injection to animals, the major route of entry of the product into the soil environment is probably via excretion following use and the subsequent disposal of contaminated manure onto land (Halling-Sørensen et al. 2001; Boxall et al. 2004). Drugs administered to grazing animals or animals reared intensively outdoors may be deposited directly to land or surface water in dung or urine, exposing soil organisms to high local concentrations (Sommer et al. 1992; Halling-Sørensen et al. 1998; Montforts 1999; Floate et al. 2005).

The fate and subsequent transport of a given medicine in soil will depend on its specific physical and chemical properties, as well as site-specific climate conditions that are rate limiting for biodegradation (e.g., temperature) and soil characteristics (e.g., pH, organic matter, or clay content) that determine availability for transport and for biodegradation. For example, the propensity for sorption to soil organic matter (the K_{oc}) will influence the potential for mobility through leaching. Overall, knowledge of soil physical and chemical properties combined with data from environmental fate studies will confirm if a substance is classified as biodegradable, persistent, or a risk to other compartments (e.g., surface water or groundwater).

In this chapter, we describe those factors and processes determining the inputs and fate of veterinary medicines in the soil environment. Models used for estimating concentrations of veterinary medicines in animal manure and in soil, and the fate and behavior of these medicines once in the terrestrial environment, are also described. We conclude by identifying a number of knowledge gaps that should form the basis for future research.

6.2 ABSORPTION AND EXCRETION BY ANIMALS

Knowledge about the kinetics of the veterinary medicine after application to the target animals is of tremendous relevance within the development of a veterinary medicinal product. This is obtained from the adsorption, distribution, metabolism, and excretion (ADME) study, which is usually undertaken with a radiolabeled parent compound. As indicated in Chapter 2, the degree of adsorption will vary with the method of application and can range from a few percent to 100%. Once absorbed the active ingredient may undergo metabolism. These reactions may result in glucuronide or sulfate conjugates or may produce other polar metabolites that are excreted in the urine or feces. The parent compound may also be excreted unchanged, and, consequently, animal feces may contain a mixture of the parent compound and metabolites. A general classification of the degree of metabolism for different types of veterinary medicine is given in Table 6.1. General assumptions may be revised where detailed ADME investigations are available (Halley et al. 1989a). ADME investigations may also provide information on the excretion of a parent compound, the amount and nature of excreted metabolites, and how these vary with application method. Metabolism data will help to identify whether the parent compound is the correct substance for further environmental assessment, or whether a major metabolite, already formed in and excreted by the animal, should be the relevant one for assessment (e.g., pro-drugs).

The formulation of veterinary medicines (e.g., aqueous or nonaqueous), the dosage, and the route of administration are key factors in determining the elimination profile for a substance. Animals tend to be treated by injection (subcutaneously or by intramuscular injection), via the feed or water, topically (as a pour-on, spot-on, or sheep dip application), by oral drench, or via a bolus releasing the

TABLE 6.1
General trend for the degree of metabolism of major therapeutic classes of veterinary medicines

Therapeutic class	Chemical group	Metabolism
Antimicrobials	Tetracyclines	Minimal
	Potentiated sulphonamides	High
	Macrolides	Minimal
	Aminoglycosides	Minimal–high
	Lincosamides	Moderate
	Fluoroquinolones	Minimal–high
Endoparasiticides — wormers	Azoles	Moderate
Endoparasiticides — wormers	Macrolide endectins	Minimal–moderate
Endoparasiticides — antiprotozoals	—	Minimal–high
Endectocides	Macrocyclic lactones	Minimal–high

Note: Classification: minimal (< 20%), moderate (20% to 80%), high (> 80%).
Source: Classification taken from Boxall et al. (2004).

TABLE 6.2
Parasiticide formulations available in the United Kingdom

Parasiticide	Cattle	Sheep
Albendazole	Oral	Oral
Cypermethrin	—	Dip
Deltamethrin	Pour-on	
Spot-on	Spot-on	
Diazinon	—	Dip
Doramectin	Subcutaneous injection	Intramuscular injection
Eprinomectin	Pour-on	—
Fenbendazole	Oral suspension	
Oral bolus		
Feed	Oral suspension	
Ivermectin	Injection	
Pour-on	Injection	
Oral		
Levamisole	Oral	
Pour-on	Oral	
Morantel	Bolus	—
Moxidectin	Injectable	
Pour-on	Injectable	
Oral drench		
Oxfendazole	Pulse release bolus	
Oral	Oral	
Triclabendazole	—	Oral

Source: National Office of Animal Health (2007).

drug over a period of time. Many medicines commonly used are available in one or more application types and formulations (e.g., Table 6.2). For example, fenbendazole is available in the United Kingdom as an oral drench for cattle and sheep at different concentrations and as a bolus for cattle, continuously releasing fenbendazole for 140 days.

Pour-on treatments result in higher and more variable concentrations than injectable treatments, and compounds are excreted more rapidly following oral applications. Most studies on this in the literature concern the different methods of administering ivermectin. Herd et al. (1996) investigated the effect of 3 ivermectin application methods upon residue levels excreted in cattle dung over time (Figure 6.1). Ivermectin residues following a pour-on application resulted in a higher initial peak of 17.1 mg kg^{-1} (dry weight) occurring 2 days after treatment. Comparable results were obtained by Sommer and Steffansen (1993), where peak excretion of 9 mg kg^{-1} (dry weight) occurred 1 day after pour-on. Subcutaneous injection was found to result in a slightly later and considerably lower peak excretion of 1.38 mg kg^{-1} (dry weight) after 3 days by Herd et al. (1996). Sommer and

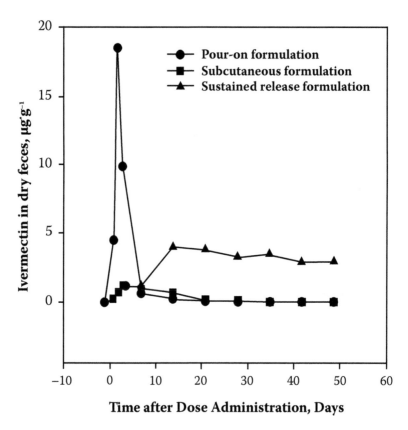

FIGURE 6.1 Excretion profiles of ivermectin following 3 different application methods. *Source*: Reprinted from Intl J Parasitol 26(10), Herd RP, Sams RA, Ashcraft SM, Persistence of ivermectin in plasma and feces following treatment of cows with ivermectin sustained release, pour-on or injectable formulations, 1087–1093 (1996), with permission from Elsevier.

Steffansen (1993) reported a peak of 3.9 mg kg^{-1} (dry weight) after 2 days. After approximately 5 days, both studies found that both pour-on and injection residue levels declined at a similar rate. Sommer et al. (1992) provide an example of how the considerations above can affect exposure for ivermectin applied to cattle by subcutaneous or topical (pour-on) application. Maximum excretion concentration (C_{max}) may differ by at least a factor of 2. In Sommer et al.'s (1992) data, values of 4.4 ppm versus 9.6 ppm were obtained. The value for t_{max} (the time to the maximum excretion concentration) may also be slightly different due to absorption and distribution processes, whereas the overall time of excretion of relevant amounts may be similar.

Differences in peak excretion levels between pour-on and injectable ivermectin formulations (e.g., Figure 6.1) were attributed to a slower release from the subcutaneous depot, rapid absorbance through the skin, and differences in the dose rate (Herd et al. 1996). However, Laffont et al. (2003) found the major route of ivermectin absorbance after pour-on to be oral ingestion after licking, and not absorbance through the skin (accounting for 58% to 87% and 10% of the applied dose, respectively). This led to high variability (between and within animals) in fecal excretion, and, in addition, most of the applied dose was transmitted directly to the feces. Doramectin and moxidectin were also found to be transferred via licking to untreated cattle (Bousquet-Melou et al. 2004). It would therefore appear that fecal residues of veterinary medicines following pour-on application are more difficult to predict than is the case for other forms of application.

Several studies have indicated that residues are excreted more rapidly following oral (aqueous) treatment compared to injectable (nonaqueous) treatments. When comparing both treatments to sheep, Borgsteede (1993) demonstrated that the injectable formulation of ivermectin had a longer resident time in sheep than the oral formulation. Wardhaugh and Mahon (1998) found that dung from cattle treated with injectable ivermectin remained toxic to dung containing dung-breeding fauna for a longer period of time compared to dung from orally treated cattle. As the two treatments were of the same dose, it was concluded that the oral formulation is eliminated more rapidly than the injectable formulation. The pattern of excretion following treatment using a bolus is clearly very different. Boluses are designed to release veterinary medicines over a prolonged period of time, as either a pulsed or sustained release. Following use of the sustained-release bolus, Herd et al. (1996) found that fecal ivermectin levels remained relatively constant at a mean of 0.4 to 0.5 mg kg^{-1} (dry weight) from approximately 14 days after application to the end of the study.

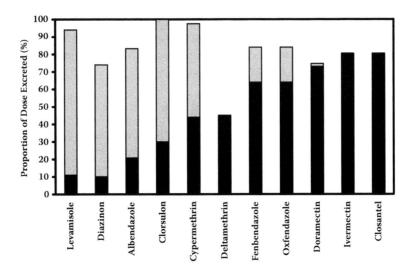

FIGURE 6.2   The percentage of the applied dose excreted in the dung (in black) and urine (in gray), as parent molecule and/or metabolites. *Source*: Inchem (1993), European Agency for the Evaluation of Medicinal Products (1999), Inchem (2006), Hennessy et al. (2000); Hennessy et al. (1993b); Paulson and Feil (1996); Hennessy et al. (1993a); Juliet et al. (2001); Croucher et al. (1985).

After application the active ingredient may be excreted as the parent compound and/or metabolites in the feces or urine of the animal. Figure 6.2 shows the proportion of the applied dose excreted in the dung or urine for a range of parasiticides used in the United Kingdom for pasture animals. The avermectins as a group (e.g., ivermectin and doramectin) tend to be excreted in the feces, with only a small proportion of the applied dose detected in the urine (Chiu et al. 1990; Hennessy et al. 2000). However, there appears to be a large variation in the excretion route of the benzimidazoles, with the applied dose of albendazole and oxfendazole largely excreted in the urine and feces, respectively (Hennessy et al. 1993a, 1993b).

Veterinary medicines excreted in urine tend to be extensively metabolized. For example, when animals are treated orally with levamisole a large proportion of the applied dose is detected in the urine, whereas the parent molecule is not (Paulson and Feil, 1996). Diazinon is also readily metabolized, with 73% to 81% of the applied dose excreted in the urine, and less than 1% present as diazinon (Inchem 1970). Veterinary medicines excreted via feces tend to contain large proportions of the unchanged parent molecule. For example, a large proportion of applied radiolabeled ivermectin (39% to 45%) was excreted in feces as the parent compound (Halley et al. 1989a). In addition, 86% of the fecal residues of eprinomectin (closely related to ivermectin) were parent compound (Inchem 1998). Closantel is also poorly metabolized, with 80% to 90% of the fecal residues excreted as unchanged closantel (Inchem 2006).

Residue data in target (food-producing) animals used to define withdrawal periods may also be used to give an indication of the potential for bioaccumulation in the environment. However, it must be noted that the compound under consideration should be the same as that for which the withdrawal data are generated and also be of relevance in the environment. Long withdrawal periods of several weeks may indicate such a potential for accumulation.

6.3 FATE DURING MANURE STORAGE

For housed animals, the veterinary medicine will be excreted in the feces or urine, and these will then be collected and stored prior to use as a fertilizer. During the storage period, it is possible that the veterinary medicines will be degraded. No validated or standardized method for assessing the fate of veterinary medicines in manure at either the laboratory or field level exists, and tests in existing pesticide or OECD guidelines do not cover these aspects. In many confined animal and poultry production systems, waste is stored for some time, during which a transformation of veterinary medicines could occur prior to release of material into the broader environment. Various production systems typically store waste as a slurry; others store it as a solid (Table 6.3). Factors that control dissipation rates and pathways such as temperature, redox conditions, organic matter content, and pH will vary widely according to the storage method employed and climatic conditions. Manure-handling practices that could accelerate veterinary medicine

TABLE 6.3
Commonly employed practices for manure storage and handling

System	Manure stored as	Treatment options[a]
Poultry broiler	Solid (mixing with bedding)	Composting
Poultry layer	Slurry	Static storage, aeration
Beef	Solid	Composting
Dairy	Slurry	Static storage, anaerobic digestion
Swine	Slurry	Static storage, aeration, composting, anaerobic digestion

[a] Fecal material will typically be mixed with some bulking agent (e.g., straw or sawdust) prior to composting. Stored slurry can be aerated by pumped-in air or passively with wind-driven turbines (e.g., Pondmill). Both aerobic composting and anaerobic digestion (for biogas production) will result in increased temperature.

dissipation (e.g., composting) offer an opportunity to reduce environmental exposure significantly.

When testing the fate of a veterinary medicine in manure or slurry, the choice of the test matrix will depend upon the proposed treatment group of the compound (e.g., cattle, pig, or poultry). The matrix is less likely to influence the degradation pathway than the conditions (aerobic or anaerobic); therefore, an aerobic study in cattle manure is an acceptable surrogate for an aerobic study in pig or poultry litter, although the moisture content could be an influencing factor for some compounds.

It is important to consider the measured concentrations of veterinary medicines in the manure, manure type, storage conditions in the tank, mode of medication, agricultural practice, solids concentration, organic carbon concentration, water content, pH, temperature, and redox conditions in different layers of the tank, as all these factors can influence the degradation process. Degradation may also be influenced under methanogenic, denitrifying, and aerobic conditions. The deconjugation rate of excreted veterinary medicines in manure may be significant and require further study under the relevant conditions.

Laboratory degradation studies of active substances in soil may not be sufficient to predict degradation rates in dung and manure (Erzen et al. 2005). Data are available on the persistence in manure of a range of commonly used classes of antibiotic veterinary medicines (reviewed in Boxall et al. 2004). Sulfonamides, aminoglycosides, beta-lactams, and macrolides have half-lives of 30 days or lower and are therefore likely to be significantly degraded during manure and slurry storage (although no data are available on the fate of the degradation products). In contrast, the macrolide endectin, ivermectin, tetracyclines, and quinolones have longer half-lives and are therefore likely to be more persistent. Results giving degradation rate coefficients of the different veterinary medicines in manure are not necessarily related to agricultural practice when handling manure, although degradation rates in manure are generally faster than those in soil. For example, under methanogenic conditions the degradation half-life for tylosin A was less than 2 days (Loke et al. 2000). We recommend that systematic experimental determination of veterinary medicine persistence in appropriate manures incubated under realistic conditions should be performed.

6.4   RELEASES TO THE ENVIRONMENT

For housed animals, the main route of release of veterinary medicines to the soil environment will be via the application of manure or slurry to soils as a fertilizer. In most jurisdictions, regulations and guidelines that mandate manure application practices are based on crop nitrogen or phosphorus needs and site-specific considerations, including climate and land characteristics. Manure application rates, manure application timing, manure incorporation into soil, suitable slope, and setback (buffer) distances from surface water may be specified or required. These best management practices (BMPs) are designed to protect adjacent water resources from contamination with enteric bacteria or nutrients. It remains to be determined if these practices are suitably protective of exposure from veterinary medicines. The characteristics of these practices are summarized in Table 6.4.

Although inputs from housed, intensively reared animal facilities tend to be considered the worst case in terms of environmental exposure, in some instances the pasture situation may be of more concern, particularly when considering

---

TABLE 6.4

Characteristics of manure type or application of best management practices (BMP) that can influence the persistence of veterinary medicines in soil

Factor	Features influencing persistence
Manure type	
Solid	Heterogeneity of application and poor soil contact, diffusivity of oxygen
Slurry	Immediate contact with soil, moisture available for microbial activity, risk of off-site movement
Chicken litter	Heterogeneity of application, high proportion of cellulolytic material (straw, wood shavings, sawdust)
Application method	
Broadcast (surface application)	Poor contact with soil, dessication, exposure to sunlight, risk of off-site movement
Broadcast (incorporated)	Good contact with soil, lower risk of off-site movement
Injection	Good contact with soil, lower risk of off-site movement
Cropping	
Standing crop	Rhizosphere stimulation of biodegradation
Bare soil	Evapotranspiration moisture reduction

potential effects on dung fauna. Compounds in manure stored prior to application to the land will have the opportunity to undergo anaerobic degradation, whereas veterinary medicines given to grazing animals will usually be excreted directly to the land.

The presence of parasiticide residues in the pasture environment will depend on a number of factors including method of medicine application, degree of metabolism, route of excretion (via urine or feces), and persistence in the field. In addition, at the larger scale, factors such as treatment regime, stocking density, and proportion of animals treated will also influence concentrations in the field. The following sections discuss the factors that influence the likely concentration of veterinary medicine residues.

6.5 FACTORS AFFECTING DISSIPATION IN THE FARM ENVIRONMENT

"Dissipation" as originally defined for pesticides is the decrease in extractable pesticide concentration due to transformation (both biological and chemical) and the formation of nonextractable or "bound" residues with the soil (Calderbank 1989). The same definition is used here for veterinary medicines. In the following sections, we describe those factors and processes affecting dissipation in dung and soil systems.

6.5.1 DISSIPATION AND TRANSPORT IN DUNG SYSTEMS

For pasture animals, once excreted, veterinary medicines and their metabolites may break down or persist in the dung on the pasture. Drug residues in dung may be subject to biodegradation, leaching into the soil, or photodegradation, or be physically incorporated into the soil by soil organisms. Persistence of residues in the field will be heavily influenced by climatic conditions. Differences in location and season will affect both chemical degradation and dung degradation. Results from studies of avermectin persistence in the field ranged from no degradation at the end of a 180-day study in Argentina to complete degradation after 6 days (Lumaret et al., 1993; Suarez et al., 2003). In laboratory studies there is also enormous variation in the degradation rate with soil type and the presence or absence of manure (Bull et al. 1984; Halley et al. 1989a, 1989b; Lumaret et al. 1993; Sommer and Steffansen, 1993; Suarez et al. 2003; Erzen et al. 2005). Mckellar et al. (1993) reported consistently lower morantel concentrations in the crust of cow pats compared to the core over 100 days, suggesting that surface residues were subject to photolysis. However, as there is little exposure to sunlight within the dung pat, this was judged unlikely to present a significant route of degradation overall.

At the field scale, the residence time in the field and the overall concentration of veterinary medicines in dung will be affected by a number of factors, including frequency of treatments in a season, stocking density, and the proportion of animals treated. Pasture animals may be treated with veterinary medicines at different times during the grazing season and at different frequencies. For example, the recommended dosing for cattle using doramectin in Dectomax injectable formulation is once at turnout (around May in the United Kingdom) and again 8 weeks later (National Office of Animal Health [NOAH] 2007). Ivomec classic, a pour-on containing ivermectin, recommends treating calves 3, 8, and 13 weeks after the first day of turnout (NOAH 2007). However, the moxidectin treatment used in Cydectin pour-on for cattle may be used for late grazing in September or just prior to rehousing. In addition, in some circumstances not the entire herd of animals is treated with veterinary medicines. A recent survey of the use of parasiticides in cattle farms in the United Kingdom found that the proportion of dairy and beef cattle treated with parasiticide varied from 10% to 100%, although it was rare that the entire herd was treated at the same time (Boxall et al., 2007). The same survey also found that the majority of farmers separated their treated and untreated cattle when they were released to pasture.

Persistence of residues will be heavily influenced by climatic conditions, differing between location and season and affecting chemical degradation and dung degradation. For example, Halley et al. (1989a) found that the degradation of ivermectin would be in the order of 7 to 14 days under summer conditions and in the order of 91 to 217 days in winter. The timing of application of manure or slurry to land may therefore be a significant factor in determining the subsequent degradation rate of a compound.

6.5.2 DISSIPATION AND TRANSPORT IN SOIL SYSTEMS

When a veterinary medicine reaches the soil, it may partition to the soil particles, run off to surface water, leach to groundwater, or be degraded. Over time most compounds dissipate from the topsoil. The dissipation of veterinary drugs in soil has been the topic in a number of studies (e.g., Blackwell et al. 2007; Halling-Sørensen et al. 2005). The dissipation of veterinary antibiotics following application to soil can be variously due to biodegradation in soil or soil–manure mixtures, chemical hydrolysis, sequestration in the soil due to various sorptive processes, or transport to another environmental compartment.

6.5.2.1 Biotic Degradation Processes

The main mechanism for dissipation of veterinary medicines in soils is via aerobic biodegradation. Degradation rates in soil vary, with half-lives ranging from days to years (reviewed in Boxall et al. 2004; and see Table 6.5). Degradation of veterinary medicines is affected by environmental conditions such as temperature and pH and the presence of specific degrading bacteria that have developed to degrade groups of medicines (Gilbertson et al. 1990; Ingerslev and Halling-Sørensen 2001). As well as varying significantly between chemical classes, degradation rates for veterinary medicines also vary within a chemical class. For instance, of the quinolones, olaquindox can be considered to be only slightly persistent (with a half-life of 6 to 9 days), whereas danofloxacin is very persistent (half-life 87 to

TABLE 6.5
Mobility and persistence classifications for a range of active ingredients used in veterinary products

	Nonpersistent (DT$_{50}$ < 5 days)	Slightly persistent (DT$_{50}$ 5–21 days)	Moderately persistent (DT$_{50}$ 22–60 days)	Very persistent (DT$_{50}$ > 60 days)	Unknown
Very mobile (K_{oc} < 15)	Sulfamethazine				
Mobile (K_{oc} 15–74)		Metronidazole	Clorsulon Forfenicol		
Moderately mobile (K_{oc} 75–499)	Sulfadimethoxine	Olaquindox Piperonyl butoxide	Ceftiofur	Chlorfenvinphos Diclazuril (silty clay loam)	
Slightly mobile (K_{oc} 500–4000)	Tylosin (soil and manure)	Diazinon Tylosin (soil only) Emamectin benzoate		Eprinomectin Diclazuril (sandy loam and silt loam) Oxfendazole	Efrotomycin (loam, silt loam)
Nonmobile (K_{oc} > 4000)		Avermectin B1a (sandy loam soil)	Avermectin B1a (sandy soil) Deltamethrin	Albendazole Coumaphos Cypermethrin Danofloxacin Doramectin Erythromycin Ivermectin Moxidectin Oxytetracycline Selamectin Sarafloxacin	Ciprofloxacin Efrotomycin (sandy loam, clay loam) Enrofloxacin Ofloxacin Tetracycline
Unknown K_{oc}					

Source: Hollis (1991).

143 days). In addition, published data for some individual compounds show that persistence varies according to soil type and conditions. In particular, diazinon was shown to be relatively labile (half-life 1.7 days) in a flooded soil that had been previously treated with the compound, but was reported to be very persistent in sandy soils (half-life 88 to 112 days) (Lewis et al. 1993). Of the available data, coumaphos and emamectin benzoate were the most persistent compounds in soil (with half-lives of 300 and 427 days, respectively), whereas tylosin and dichlorvos were the least persistent (with half-lives of 3 to 8 days and < 1 day, respectively).

A number of suitable validated guideline methods developed for pesticide scenarios exist for examining degradation under aerobic, anaerobic, and denitrifying conditions. These may be a starting point for assessing veterinary medicines. An important question also to consider is the role of manure in soil systems in terms of degradation pathways and removal rates.

Manure amendment changes the properties of the soil system by increasing water content and organic carbon and by modifying pH and the buffering capacity of the soil. Furthermore, inclusion of manure alters bacterial abundance and diversity in the topsoil. Whether changes in microbiological degradation pathways result from manure inclusion is not currently known. Initial laboratory-scale investigations suggest that manure inclusion up to 10% by weight does not affect the rate of degradation of tylosin, olaquindox, and metronidazole (Ingerslev and Halling-Sørensen 2001). But recent studies have shown that when manure is combined with soil, degradation may be enhanced for selected medicines such as sulfadimethoxine (Wang et al. 2006).

Compounds can be applied to the field in solid or slurried manure, with either a surface or subsurface application. No guidance exists on the methods to be used to evaluate veterinary medicine degradation in the field, but the practices employed in pesticide field dissipation studies may be used in this context, as the scenarios are very similar. It is important that the application method selected reflects common agronomic practice for the situation under consideration. Assessing antibacterial and fungicidal agents at unrealistically high spiking levels of the compounds may give false data on biotic removal due to bacteriostatic or bacteriocidal effects of tested compounds. Radiolabeled antimicrobial agents may also not be commercially available as they can be difficult to produce due to their semisynthetic origin.

Few studies have been carried out in the field, so limited data are available on veterinary medicine field dissipation (Kay et al. 2004; Halling-Sørensen et al. 2005; Blackwell et al. 2007).

6.5.2.2 Abiotic Degradation Processes

Depending on the nature of the chemical, other degradation and depletion mechanisms may occur, including soil photolysis, hydrolysis, and soil complex formation. The degradation products of both photolytic and hydrolytic degradation processes may undergo aerobic biodegradation in upper soil layers or anaerobic degradation in deeper soil layers. For many medicines, both hydrolysis and photolysis may be important dissipation pathways. Once manure is incorporated into the soil these processes are less important, but they may still be relevant in water. ISO, OECD, and other standardizing bodies have developed appropriate methods for chemical substances for assessing hydrolysis, photolysis, and soil sorption. However, once again the influence of manure amendment should be considered for veterinary medicines, if appropriate.

6.5.2.3 Sorption to Soil

The degree to which veterinary medicines may adsorb to particulates varies considerably (Table 6.5), and this also affects the potential mobility of the compound. This can be influenced by the pH of the soil, depending on the ionic state of the compound under consideration. Partition coefficients (K_D) range from low (0.6 L kg^{-1}) to high (6000 L kg^{-1}) adsorption (K_{oc}; the organic normalized partition coefficient ranges from 40 to 1.63×10^7 L kg^{-1}). In addition, the variation in partitioning for a given compound in different soils can be significant (up to a factor of 30 for efrotomycin).

The range of partitioning values can be explained to some extent by studies addressing the sorption of tetracycline and enrofloxacin. The results suggest that surface interactions of these compounds with clay minerals are responsible for the strong sorption to soils. The underlying processes are cation exchange (tetracycline at low pH) and surface complex formation with divalent cations sorbed at the clay surfaces (tetracycline at intermediate pH and enrofloxacin at high pH). This indicates that in order to arrive at a realistic assessment of the availability of these compounds for transport through the soil and uptake into soil organisms, soil chemistry may not be reduced to the organic carbon content but the clay content, the pH of the soil solution, and the coverage of the ion exchange sites need to be accounted for.

Manure and slurry may also alter the behavior and transport of veterinary medicines. Studies have demonstrated that the addition of these matrices can affect the sorption behavior of veterinary medicines and that they may affect persistence (Boxall et al. 2002; Thiele-Bruhn and Aust 2004). These effects have been attributed to changes in pH or the nature of dissolved organic carbon in the soil and manure system.

Guideline methods applicable to veterinary medicines are published by several regulatory bodies (e.g., the ISO and OECD). A substantial number of published data on sorption coefficients can be found in the open literature and are often higher than expected from their lipophilicity (e.g., tetracyclines and quinolones; Tolls 2001). Thus quantitative structure-activity relationships based on parameters such as K_{ow} can overestimate mobility. Coefficients are concentration dependent, and high spiking concentrations may give unrealistic results.

6.5.3 Bound Residues

Nonextractable residues are formed in soils during the application of pesticides (Führ 1987; Calderbank 1989). Sequestered residues have the potential to be transported to subsurface water through preferential flow. More detailed experiments are needed to understand these mechanisms for veterinary medicines, and the VICH guidelines indicate that a case-by-case evaluation has to be conducted. The ionic nature of veterinary medicines makes it difficult to predict their behavior under all conditions. Time-dependent sorption appears to be a very important mechanism of removal for certain compounds (e.g., tetracyclines). Bound residues are also an important aspect in effect studies and are dealt with in Chapter 7 of this book.

The mechanisms by which residues become bound are numerous and relate to both the target molecule and the specific soil type. Characterization of bound residues by extraction with organic solvents, treatment with acid–base reflux procedures, and enzymes may assist in defining the fraction of the soil to which the residue is associated. However, these procedures can only be effectively conducted where the parent compound was applied in a radiolabeled form, and such analyses will not necessarily provide information on the structure of the residues released. Residues from biomass or highly degraded compounds are not considered bound residues by the International Union of Pure and Applied Chemistry (IUPAC) definition of pesticides (Roberts 1984). However, bound residues cannot be distinguished from biogenic residues, because the chemical structures of the residues are not known. The chemical reactivity of an active compound or of a metabolite governs the formation of bound residues, whose levels may range from 7% to 90% of the quantity applied (Calderbank 1989). Many pesticides are partially degraded, and the metabolites are involved in the formation of bound residues (Hsu and Bartha 1976).

Only a few studies have addressed the question of bound residues of veterinary medicines. Chander et al. (2005) investigated the process by sorbing various amounts of tetracycline or tylosin on two different textured soils (Webster clay loam [fine-loamy, mixed, superactive, mesic Typic Endoaquolls] and Hubbard loamy sand [sandy, mixed, frigid Entic Hapludolls]), incubating these soils with three different bacterial cultures (an antibiotic-resistant strain of *Salmonella* sp. [*Salmonella*[R]], an antibiotic-sensitive strain of *Salmonella* sp. [*Salmonella*[S]], and *Escherichia coli* ATCC 25922), and then enumerating the number of colony-forming units relative to the control. Soil-adsorbed antibiotics were found to retain their antimicrobial properties because both antibiotics inhibited the growth of all three bacterial species. Averaged over all other factors, soil-adsorbed antimicrobial activity was higher for Hubbard loamy sand than for Webster clay loam, most likely due to the higher affinity (higher clay content) of the Webster soil for antibiotics. Similarly, there was a greater decline in bacterial growth with tetracycline than with tylosin, likely due to greater amounts of soil-adsorbed tetracycline and also due to the lower minimum inhibitory concentration of most bacteria for tetracycline compared with tylosin. The antimicrobial effect of tetracycline was also greater under dynamic than static growth conditions, possibly because agitation under dynamic growth conditions helped increase tetracycline desorption and/or increase contact between soil-adsorbed tetracycline and bacteria. Chander et al. (2005) concluded that even though antibiotics are tightly adsorbed by clay particles, they are still biologically active and may influence the selection of antibiotic-resistant bacteria in the terrestrial environment.

6.6   UPTAKE BY PLANTS

The potential for medicines to be taken up by plants has also been considered (e.g., Migliore et al. 1996, 1998, 2000; Forni et al. 2001, 2002; Kumar et al. 2005; Boxall et al. 2006). Uptake of fluoroquinolones, sulfonamides, levamisole, trimethoprim, diazinon, chlortetracycline, and florfenicol has been demonstrated experimentally. Uptake can differ according to the crop type. For example, Boxall et al. (2006) demonstrated that florfenicol, levamisole, and trimethoprim were taken up by lettuce, whereas diazinon, enrofloxacin, florfenicol, and trimethoprim were detected in carrot roots. Kumar et al. (2005) showed in a greenhouse study in which manure was applied to soil that the plants absorbed antibiotics present in the manure. The test crops were corn (*Zea mays*), green onion (*Allium cepa*), and cabbage (*Brassica oleracea*). All three crops absorbed chlortetracycline but not tylosin. The concentrations of chlortetracycline in plant tissues were small (2 to 17 ng g^{-1} fresh weight), but these concentrations increased with increasing amounts of antibiotics present in the manure. Such studies point out the potential risks to humans and wildlife associated with consumption of plants grown in soil amended with antibiotic-laden manures.

6.7   MODELS FOR ESTIMATING THE CONCENTRATION OF VETERINARY MEDICINE IN SOIL

From the above, it is clear that the exposure of the environment to a veterinary medicinal product is determined by a range of factors and processes. When assessing the environmental risks posed by a new product, models and model scenarios are typically used to estimate the level of exposure. For environmental risk assessment purposes, these modeling approaches must be responsive to regional soil and climate conditions, as well as manure storage and handling conditions that can influence the persistence of excreted residues. Regional agronomic considerations and regulations that proscribe and constrain manure application rates, timing, and method must likewise be considered. Some emission scenarios (e.g., sheep dipping) are very country or even region specific. Currently employed terrestrial assessment models generally assume that residues, following excretion, are uniformly distributed in the terrestrial environment. In fact the distribution may be quite patchy, particularly in the case of dung that is excreted by animals on pasture. Currently, terrestrial exposure assessments contain the following elements:

- Information on the treatment of terrestrial animals
- Factors influencing the uptake and excretion of veterinary medicines by the animals
- Factors affecting how much residue reaches the land
- Factors affecting dissipation once the substance reaches the soil

In the following sections, we describe these models in more detail.

6.7.1 INTENSIVELY REARED ANIMALS

For intensively reared animals that are housed indoors throughout the production cycle, treatment with the veterinary medicine is carried out in housed animals, and the active residue is excreted indoors and incorporated into the slurry or farmyard manure. This active residue reaches the environment when the manure from the stable is spread onto land. A number of models have been proposed to enable the calculation of the concentration of a veterinary medicine in soil after spreading manure from treated animals, based on a fixed amount of manure that can be spread on an area of land, and then incorporation to a uniform depth of soil. The mass of manure spread per unit area is usually controlled by the amount of nitrogen or, less frequently, by the amount of phosphorus in the manure.

The first of these methods was developed by Spaepen et al. (1997). In this method the concentration of the veterinary medicine in manure is calculated after treatment of the housed animals. In addition to the dose and duration of treatment, the calculation requires information on the body weight of the individual animal at treatment, the number of animals kept in 1 stable or barn each year, and the annual output of manure from the stabled animal. Following calculation of the concentration of veterinary medicine in manure, the quantity of manure that is spread per hectare of land is determined. The rate is controlled by the nitrogen or phosphorus content of the manure, which is provided in the publication with default values for most of the other parameters. The PEC$_{soil}$ is calculated by calculating the mass of veterinary medicine spread per hectare of soil divided by the weight of the soil in the layer into which the residue penetrated, plus the weight of the manure (Equations 6.1 to 6.4). The PEC$_{soil}$ is an annual value. An evaluation of this method against measured concentrations for veterinary medicines in the field indicates that it is likely to produce conservative exposure estimates (Blackwell et al. 2005).

$$M = D \times BW \times T \times C \tag{6.1}$$

$$C_{excreta} = \frac{M}{P_{excreta}} \tag{6.2}$$

$$R_{hectare} = C_{excreta} \times \frac{170}{N_{prod}} \times P_{excreta} \tag{6.3}$$

$$PEC_{soil} = \frac{R_{hectare} \times 1000}{\left(\frac{5}{100} \times 1500 \times 10000\right) + \left(\frac{170}{N_{prod}} \times P_{excreta}\right)} \tag{6.4}$$

where
PEC_{soil} = predicted environmental concentration in soil (µg kg^{-1})
    M = total dose administered (mg)
    D = dosage used (mg kg^{-1} body weight d^{-1})
    T = number of daily administrations in 1 course of treatment (days)
    BW = animal body weight (kg)
    C = number of animals raised per place per year
    C_{excreta} = concentration of active ingredient in excreta (mg kg^{-1})
    P_{excreta} = excreta produced per place per year (kg y^{-1})
    N_{prod} = nitrogen produced per place per year (kg N y^{-1})
    1500 = soil bulk density (kg m^{-3})
10000 = area of 1 hectare (m^2 ha^{-1})
        5 = depth of penetration into soil (cm)
    R_{hectare} = mass of active spread per hectare (mg ha^{-1})
    1000 = conversion factor (µg kg^{-1})

A similar method to calculate the PEC$_{\text{soil}}$ was developed by the Animal Health Institute (AHI) and Center for Veterinary Medicine (CVM) in the United States (Robinson personal communication 2006). In this method the concentration of the drug in manure is calculated by multiplying the dose per animal (mg kg^{-1} body weight) by the number of treatments and dividing by the total amount of manure produced in the production period. The PEC$_{\text{soil}}$ is calculated by multiplying the concentration of the drug in manure by the amount of manure allowed to be spread per hectare (a fixed value for each of cattle, pigs, and poultry) and dividing by the mass of 1 hectare of soil mixed to a depth of 15 cm. The value is an annual value.

Montforts (1999) developed a method specifically for the situation in the Netherlands, where the quantity of manure that can be spread onto land is restricted by the amount of phosphorus allowed.

The method of Montforts and Tarazona (2003) assumes that the average storage time for manure on the farm before spreading is 30 days. It is assumed that the treatment of the animals with the product occurs during the 30-day storage period and then the manure is spread onto land to comply with the nitrogen standard. This method does not consider the number of animals kept per stable unit per year (Equation 6.5).

$$\text{PEC}_{\text{soil}} = \left(\frac{D \times T \times BW \times 170}{1500 \times 10000 \times 0.05 \times N} \right) \times 1000 \qquad (6.5)$$

where
PEC_{soil} = predicted environmental concentration in soil (µg kg^{-1})
    D = dosage used (mg kg^{-1} body weight d^{-1})

T = number of daily administrations in 1 course of treatment (days)
BW = animal body weight (kg)
170 = EU nitrogen-spreading limit (kg N ha^{-1} y^{-1})
1500 = soil bulk density (kg m^{-3})
10000 = area of 1 hectare (m^2 ha^{-1})
0.05 = depth of penetration into soil (m)
N = nitrogen produced in 30 days (kg N)

A fifth method has been proposed recently in a draft guideline published for consultation by the Committee for Medicinal Products for Veterinary Use (CVMP 2006; see Equation 6.6). The method is again based on spreading manure according to the nitrogen content of the manure. The number of animals occupying a stable unit over the year is also considered.

$$\text{PEC}_{\text{soil}} = \left(\frac{D \times T \times BW \times C \times 170 \times F}{1500 \times 10000 \times 0.05 \times N \times H} \right) \times 1000 \qquad (6.6)$$

where
PEC_{soil} = predicted environmental concentration in soil (µg kg^{-1})
D = dosage used (mg kg^{-1} body weight d^{-1})
T = number of daily administrations in 1 course of treatment (days)
BW = animal body weight (kg)
C = number of animals raised per place per year
170 = EU nitrogen-spreading limit (kg N ha^{-1} y^{-1})
F = fraction of herd treated (value between 0 and 1)
1500 = soil bulk density (kg m^{-3})
10000 = area of 1 hectare (m^2 ha^{-1})
0.05 = depth of penetration into soil (m)
N = nitrogen produced in 1 year (kg N y^{-1})
H = housing factor (either one for animals housed throughout the year or 0.5 for animals housed for only 6 months)
1000 = conversion factor (µg kg^{-1})

These 5 methods of calculating a PEC$_{\text{soil}}$ value can be compared using a standard treatment scenario of a hypothetical veterinary medicine dosed at 10 mg kg^{-1} body weight for 5 days. The PEC$_{\text{soil}}$ values resulting from the different calculation methods are given in Table 6.6. In general, the PEC$_{\text{soil}}$ values calculated using the phosphorus standard to control the amount of manure spread onto land are the lowest. The method of Montforts and Tarazona (2003) gives the highest values when used to calculate the PEC for animals that have a single production cycle per year.

A comparison of predicted concentrations, obtained for the Spaepen, CVMP, and Montforts and Tarazona models, with measured environmental concentrations

TABLE 6.6
Comparison of predicted environmental concentration in soil (PEC$_{soil}$) values using different calculation methods obtained for a hypothetical veterinary medicine dosed at 10 mg kg^{-1}

	PEC$_{soil}$ value (µg kg^{-1})			
Calculation method	Fattening pig	Dairy cow	Beef bullock	Broiler
Spaepen et al. (1997)	389	69	104	877
Montforts (1999)	297	18	40	148
US AHI/CVM	692	94	45	323
Montforts and Tarazona (2003)	1228	983	1338	567
Committee for Medicinal Products for Veterinary Use (2006)	269	147	214	374

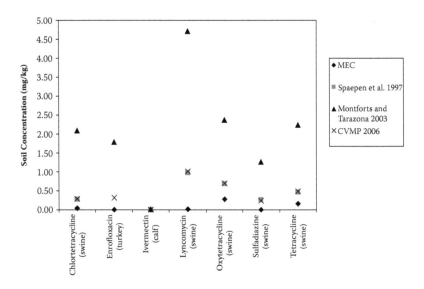

FIGURE 6.3 Measured and predicted environmental concentrations (MEC and PEC) for a range of veterinary medicines. *Source*: Measured concentrations from Hamscher et al. (2005), Boxall et al. (2006), and Zilles et al. (2005).

for a range of veterinary medicines (Figure 6.3) demonstrates that all of the models are likely to overestimate concentrations of veterinary medicines in the soil environment and that the Montforts and Tarazona (2003) model will greatly overestimate concentrations.

6.7.2 Pasture Animals

Calculation of the PEC_{soil} for pasture animals is dependent on the number of animals kept on a given area of land. This parameter is known as the stocking density and is expressed in animals per hectare. The PEC_{soil} is the total mass of active substance administered divided by a mass of soil of 750000 kg (assuming penetration to 5 cm). It is assumed that the residue is evenly distributed over the pasture. This model was proposed by the CVMP in their published draft guidance (CVMP 2006). Using the model treatment regime of 5 days of treatment of 10 mg kg^{-1} body weight, the PEC_{soil} values for dairy cattle (body weight 500 kg and stocking density 3.33 animals per hectare) and beef cattle (body weight 350 kg and stocking density 6.4 animals per hectare) are 111 µg kg^{-1} and 149 µg kg^{-1}, respectively.

In the above calculations it is assumed that the veterinary product is excreted and distributed evenly over the pasture. For many products used to treat parasites, a significant proportion of the medicine is excreted in feces. For this reason it is necessary to calculate a PEC value for the dung in order to examine the effect of this residue, in particular on dung insects. A method of calculating the PEC in dung has been proposed by the CVMP (CVMP 2006) that can be used in the absence of any excretion data, but can also be refined if excretion data are available. In this method the highest fraction of the dose excreted daily in dung (or the total dose if there is no further information) is calculated and divided by the mass of dung excreted daily. For the above example, if a single day's treatment of 10 mg kg^{-1} was excreted in feces, over the following 24 hours the PEC in dung would be 96 mg kg^{-1}, as 52 kg of dung is assumed to be excreted by a dairy cow in 24 hours.

6.7.3 PEC Refinement

The present guidelines for environmental risk assessments (especially VICH Phase II and the VICH-EU-TGD; see Chapter 3) underline the use of a "total residue approach" as the first step in estimating environmental concentrations. Under these conditions no adjustment is recommended in which available metabolism and excretion data can be used. However, exceptions may be appropriate when substantial metabolism can be demonstrated (i.e., all individual excreted metabolites are less than 5% of applied dose). In some cases it may be appropriate during the tiered risk assessment procedure to utilize metabolism data to refine PEC_{soil} or PEC_{dung}. For example, if metabolites accumulate in the animal this may reduce initial concentrations in the collected manure or the excreted dung. Consequently, after distribution of feces or manure onto land, the original PEC_{soil} can also be refined.

A different refinement may be carried out for the PEC_{dung}, dealing either with excretion data or with knowledge of which fractions are excreted via urine and which are excreted via feces. Exposure scenarios may then be refined to consider direct soil influence through urine and the residues primarily associated with dung.

6.8 RESEARCH NEEDS

Reliable methods for evaluating potential environmental exposure require both experimental data for a number of key endpoints (e.g., DT_{50} values, K_{oc}, and water solubility) as well as sophisticated modeling tools for predicting reliable and realistic environmental concentrations.

The following research needs have been identified:

- Development of clear guidance specific to veterinary medicines for laboratory and field-based methods for the evaluation of degradation and dissipation: these should take into account agronomic practice when appropriate (e.g., the addition of manure or slurry).
- Field-based validation of PEC modeling methods needs to be conducted, as there is a perception that existing methods may be too conservative and unrealistic.
- The impact of different storage and composting conditions on the degradation of veterinary medicines needs to be better understood and investigated.
- Evaluation of the potential for desorption needs to be better understood and studied.
- Exposure scenarios following the application of combination products need to be considered.

REFERENCES

Blackwell PA, Boxall ABA, Kay P, Noble H. 2005. An evaluation of a lower tier exposure assessment model for veterinary medicines. J Agric Food Chem 53(6):2192–2201.

Blackwell PA, Kay P, Boxall ABA. 2007. The dissipation and transport of veterinary antibiotics in a sandy loam soil. Chemosphere 67(2):292–299.

Borgsteede FHM. 1993. The efficacy and persistent anthelmintic effect of ivermectin in sheep. Veter Parasitol 50:117–124.

Bousquet-Melou A, Mercadier S, Alvinerie M, Toutain PL. 2004. Endectocide exchanges between grazing cattle after pour-on administration of doramectin, ivermectin and moxidectin. Intl J Parasitol 34:1299–1307.

Boxall ABA, Blackwell P, Cavallo R, Kay P, Tolls J. 2002. The sorption and transport of a sulphonamide antibiotic in soil systems. Toxicol Lett 131:19–28.

Boxall ABA, Fogg LA, Blackwell PA, Kay P, Pemberton EJ, Croxford A. 2004. Veterinary medicines in the environment. Rev Environ Contam Toxicol 180:1–91.

Boxall ABA, Johnson P, Smith EJ, Sinclair EJ, Stutt E, Levy LS. 2006. Uptake of veterinary medicines from soils into plants. J Agric Food Chem 54:2288–2297.

Boxall ABA, Sherratt TN, Pudner V, Pope LJ. 2007. A screening level index for assessing the impacts of veterinary medicines on dung flies. Environ Sci Tech 41:2630–2635.

Bull DL, Ivie GW, Macconnell JG, Gruber VF, Ku CC, Arison BH, Stevenson JM, Vandenheuvel WJA. 1984. Fate of avermectin B1a in soil and plants. J Agric Food Chem 32:94–102.

Calderbank A. 1989. The occurrence and significance of bound pesticide residues in soil. Rev Environ Contam Toxicol 108:71–101.

Chander Y, Kumar K, Goyal SM, Gupta SC. 2005. Antibacterial activity of soil-bound antibiotics. J Environ Qual 34:1952–1957.

Chien YH, Lai HT, Lui SM. 1999. Modeling the effects of sodium chloride on degradation of chloramphenicol in aquaculture pond sediment. Sci Total Environ 239:81–87.

Croucher A, Hutson DH, Stoydin G. 1985. Excretion and residues of the pyrethroid insecticide cypermethrin in lactating cows. Pestic Sci 16:287–301.

[CVMP] Committee for Medicinal Products for Veterinary Use. 2006. Committee for Medicinal Products for Veterinary Use guideline on environmental impact assessment for veterinary medicinal products in support of the VICH Guidelines GL6 and GL 38 EMEA/CVMP/ERA/418282/2005-CONSULTATION. London: CVMP.

[EMEA] European Agency for the Evaluation of Medicinal Products. 1999. EMEA Committee for Veterinary Medicinal Products: clorsulon, 1999. http://www.emea.eu.int/pdfs/vet/mrls/059099en.pdf.

Erzen NK, Kolar L, Flajs VC, Kuzner J, Marc I, Pogacnik M. 2005. Degradation of abamectin and doramectin on sheep grazed pasture. Ecotoxicol 14:627–635.

Forni C, Cascone A, Cozzolino S, Migliore L. 2001. Drugs uptake and degradation by aquatic plants as a bioremediation technique. Minerva Biotechnol 13:151–152.

Forni C, Cascone A, Fiori M, Migliore L. 2002. Sulphadimethoxine and *Azolla filiculoides Lam.*: a model for drug remediation. Water Res 36:3398–3403.

Führ F. 1987. Nonextractable pesticide residues in soil. In: Greenhalgh R, Roberts TR, editors. Pesticide science and biotechnology. Oxford (UK): Blackwell Scientific, p 381–389.

Gilbertson TJ, Hornish RE, Jaglan PS, Koshy T, Nappier JL, Stahl GL, Cazers AR, Nappier JM, Kubicek MF, Hoffman GA, Hamlow P. 1990. Environmental fate of ceftiofur sodium, a cephalosporin antibiotic: role of animal excreta in its decomposition. J Agric Food Chem 38:890–894.

Halley BA, Jacob TA, Lu AYH. 1989a. The environmental impact of the use of ivermectin: environmental effects and fate. Chemosphere 18:1543–1563.

Halley BA, Nessel RJ, Lu AYH. 1989b. Environmental aspects of ivermectin usage in livestock: general considerations. In: Campbell WC, editor. Ivermectin and abamectin. New York (NY): Springer.

Halling-Sørensen B, Jacobsen AM, Jensen J, Sengelov G, Vaclavik E, Ingerslev F. 2005. Dissipation and effects of chlortetracycline and tylosin in two agricultural soils: a field-scale study in southern Denmark. Environ Toxicol Chem 24(4):802–810.

Halling-Sørensen B, Jensen J, Tjørnelund J, Montfors MHMM. 2001. Worst-case estimations of predicted environmental soil concentrations (PEC) of selected veterinary antibiotics and residues used in Danish agriculture. In: Kümmerer K, editor. Pharmaceuticals in the environment: sources, fate, effects and risks. Berlin (Germany): Springer-Verlag.

Halling-Sørensen B, Nors Nielsen S, Lanzky PF, Ingerslev F, Holten Lützhøft HC, Jørgensen SE. 1998. Occurrence, fate and effect of pharmaceutical substances in the environment: a review. Chemosphere 36:357.

Hennessy DR, Page SW, Gottschall D. 2000. The behaviour of doramectin in the gastrointestinal tract, its secretion in bile and pharmacokinetic disposition in the peripheral circulation after oral and intravenous administration to sheep. J Veter Pharmacol Therapeut 23:203–213.

Hennessy DR, Sangster NC, Steel JW, Collins GH. 1993a. Comparative pharmacokinetic behavior of albendazole in sheep and goats. Intl J Parasitol 23:321–325.

Hennessy DR, Steel JW, Prichard RK. 1993b. Biliary-secretion and enterohepatic recycling of fenbendazole metabolites in sheep. J Veter Pharmacol Therapeut 16:132–140.

Hennessy DR, Steel JW, Prichard RK, Lacey E. 1992. The effect of coadministration of parbendazole on the disposition of oxfendazole in sheep. J Veter Pharmacol Therapeut 15:10–18.

Herd RP, Sams RA, Ashcraft SM. 1996. Persistence of ivermectin in plasma and feces following treatment of cows with ivermectin sustained-release, pour-on or injectable formulations. Intl J Parasitol 26(10):1087–1093.

Hollis JM. 1991. Mapping the vulnerability of aquifers and surface waters to pesticide contamination at the national/regional scale. In: Walker A, editor. Pesticides in soils and water: current perspectives. Proceedings of a symposium organised by the British Crop Protection Council, University of Warwick, Coventry, UK, March 25–27.

Hsu TS, Bartha R. 1976. Hydrolysable and nonhydrolysable 3,4-dichloroaniline humus complexes and their respective rates of biodegradation. J Agric Food Chem 24:118–122.

Inchem. 1970. Inchem evaluations of some pesticide residues in food: diazinon. http://www.inchem.org/documents/jmpr/jmpmono/v070pr06.htm.

Inchem. 1993. Inchem pesticide residues in food, evaluations part ii toxicology: diazinon. http://www.inchem.org/documents/jmpr/jmpmono/v93pr04.htm.

Inchem. 1998. Inchem WHO Food Additives Series 41: eprinomectin. http://www.inchem.org/documents/jecfa/jecmono/v041je02.htm.

Inchem. 2006. Inchem WHO Food Additives Series 27: closantel. http://www.inchem.org/documents/jecfa/jecmono/v27je02.htm.

Ingerslev F, Halling-Sørensen B. 2001. Biodegradability of metronidazole, olaquindox and tylosin and formation of tylosin degradation products in aerobic soil/manure slurries. Chemosphere 48:311–320.

Juliet S, Chakraborty AK, Koley KM, Mandal TK, Bhattacharyya A. 2001. Toxicokinetics, recovery efficiency and microsomal changes following administration of deltamethrin to black Bengal goats. Pest Mgmt Sci 57:311–319.

Kay P, Blackwell PA, Boxall ABA. 2004. Fate and transport of veterinary antibiotics in drained clay soils. Environ Toxicol Chem 23(5):1136–1144.

Kumar K, Gupta SC, Baidoo SK, Chander Y, Rosen CJ. 2005. Antibiotic uptake by plants from soil fertilized with animal manure. J Environ Qual 34:2082–2085.

Laffont CM, Bousquet-Melou A, Bralet D, Alvinerie M, Fink-Gremmels J, Toutain PL. 2003. A pharmacokinetic model to document the actual disposition of topical ivermectin in cattle. Veter Res 34:445–460.

Lewis S, Watson A, Hedgecott S. 1993. Proposed environmental quality standards for sheep dip chemicals in water: chlorfenvinphos, coumaphos, diazinon, fenchlorphos, flumethrin and propetamphos. WRc plc, R & D Note 216. Bristol (UK): Scotland and Northern Ireland Forum for Environmental Research and the National Rivers Authority.

Loke ML, Ingerslev F, Halling-Sørensen B, Tjørnelund J. 2000. Stability of tylosin A in manure containing test systems determined by high performance liquid chromatography. Chemosphere 40:759–765.

Lumaret JP, Galante E, Lumbreras C, Mena J, Bertrand M, Bernal JL, Cooper JF, Kadiri N, Crowe D. 1993. Field effects of ivermectin residues on dung beetles. J App Ecol 30:428–436.

Mckellar QA, Scott EW, Baxter P, Anderson LA, Bairden K. 1993. Pharmacodynamics, pharmacokinetics and fecal persistence of morantel in cattle and goats. J Veter Pharmacol Therapeut 16:87–92.

Migliore L, Brambilla G, Casoria P, Civitareale SC, Gaudio L. 1996. Effect of sulph-adimethoxine contamination on barley (*Hordeum disticum L.*, Poaceae, liliopsida). Agric Ecosyst Environ 60:121–128.

Migliore L, Civitaraele C, Cozzolino S, Casoria P, Brambilla G, Gaudio L. 1998. Laboratory models to evaluate phytotoxicity of sulphadimethoxine on terrestrial plants. Chemosphere 37:2957–2961.

Migliore L, Cozzolino S, Fiori M. 2000. Phytotoxicity to and uptake of flumequine used in intensive aquaculture on the aquatic weed, *Lythrum salicaria L.* Chemosphere 40:741–750.

Montforts MHMM. 1999. Environmental risk assessment for veterinary medicinal products, part 1. RIVM report 601300 011. Bilthoven (the Netherlands): Dutch National Institute for Public Health and the Environment (RIVM).

Montforts MHMM, Tarazona JV. 2003. Environmental risk assessment for veterinary medicinal products, part 4: exposure assessment scenarios. RIVM Report 601450017/2003. Bilthoven (The Netherlands): Dutch National Institute for Public Health and the Environment (RIVM).

National Office of Animal Health (NOAH). 2007. Compendium of data sheets for animal medicines 2007. Enfield (UK): National Office of Animal Health.

Paulson GD, Feil VJ. 1996. The disposition of C-14-levamisole in the lactating cow. Xenobiotica 26:863–875.

Roberts TR. 1984. IUPAC reports on pesticides. 17. Non-extractable pesticide-residues in soils and plants. Pure Appl Chem 56:945–956.

Robinson J. 2006. Personal communication, Kalamazoo, Michigan.

Samuelsen OB, Lunestad BT, Ervik A, Fjelde S. 1994. Stability of antibacterial agents in an artificial marine aquaculture sediment studied under laboratory conditions. Aquaculture 126:283–290.

Samuelsen OB, Lunestad BT, Husevåg B, Hølleland T, Ervik A. 1992. Residues of oxolinic acid in wild fauna following medication in fish farms. Dis Aquat Organisms 12:111–119.

Samuelsen OB, Solheim E, Lunestad BT. 1991. Fate and microbiological effects of furazolidone in a marine aquaculture sediment. Sci Total Environ 108:275–283.

Samuelsen OB, Torsvik V, Ervik A. 1992. Long-range changes in oxytetracycline concentration and bacterial resistance towards oxytetracycline in a fish farm sediment after medication. Sci Total Environ 114:25–36.

Sommer C, Steffansen B. 1993. Changes with time after treatment in the concentrations of ivermectin in fresh cow dung and in cow pats aged in the field. Veter Parasitol 48:67–73.

Sommer C, Steffansen B, Nielsen BO, Grondvold J, Jensen KMV, Jespersen JB, Springborg J, Nansen P. 1992. Ivermectin excreted in cattle dung after subcutaneous injection or pour-on treatment: concentrations and impact on dung fauna. Bull Entomol Res 82:257–264.

Spaepen KRI, Leemput LJJ, Wislocki PG, Verschueren C. 1997. A uniform procedure to estimate the predicted environmental concentration of the residues of veterinary medicines in soil. Environ Toxicol Chem 16:1977–1982.

Suarez VH, Lifschitz AL, Sallovitz JM, Lanusse CE. 2003. Effects of ivermectin and doramectin faecal residues on the invertebrate colonization of cattle dung. J App Entomol 127:481–488.

Thiele-Bruhn S, Aust MO. 2004. Effects of pig slurry on the sorption of sulfonamide antibiotics in soil. Arch Environ Contam Toxicol 47:31–39.

Tolls J. 2001. Sorption of veterinary pharmaceuticals in soils: a review. Environ Sci Technol 35:3397–3404.

Wang QQ, Bradford SA, Zheng W, Yates SR. 2006. Sulfadimethoxine degradation kinetics in manure as affected by initial concentration, moisture, and temperature. J Environ Qual 35(6): 2162–2169.

Wardhaugh KG, Mahon RJ. 1998. Comparative effects of abamectin and two formulations of ivermectin on the survival of larvae of a dung-breeding fly. Austral Veter J 76:270–272.

7 Assessing the Effects of Veterinary Medicines on the Terrestrial Environment

Katie Barrett, Kevin Floate, John Jensen,
Joe Robinson, and Neil Tolson

7.1 INTRODUCTION

This chapter summarizes, for the novice, methods used to assess risks associated with the nontarget effects of veterinary medicines in terrestrial environments. Within this broad framework, there are four specific objectives. First is to describe in general terms the functional and structural components of terrestrial ecosystems of key interest in the risk assessment process. Here, we offer suggestions on testing approaches that may vary depending upon the nature of land use. Second is to describe the existing regulatory and decision-making frameworks to assess the impacts of veterinary medicines on terrestrial ecosystems. The most widely adopted such framework was developed under the auspices of the VICH initiative (see Chapter 3), which is repeatedly referred to in the current chapter. Third is to identify the specific testing requirements for VICH phase II tiers A and B. The subsequent use of data from such tests in risk assessment is described in Chapter 3. Fourth is to identify future research needs to assess the potential risks of veterinary medicines on nontarget species in terrestrial ecosystems. Timely and accurate assessment of these potential risks benefits the regulatory authorities that are responsible for approving these products, and also the companies that market these products once approval has been granted.

7.2 CONSIDERATIONS UNIQUE TO VETERINARY MEDICINES

7.2.1 ROUTES OF ENTRY

Exposure to human medicines generally is limited to aquatic environments via entry as sewage discharge, although solid waste from sewage treatment plants is used as fertilizer in arabic situations in some countries. In contrast, veterinary medicines may enter both aquatic and terrestrial environments by several routes. In terrestrial environments, the focus of this chapter, the main route of entry occurs when stored manure accumulated from treated animals held in livestock confinements (e.g., dairies and feedlots) is spread onto land as fertilizer. Residues in manure also may be deposited directly onto pastures by treated animals. Movement of residues into terrestrial environments also may occur via disposal of waste feed or drinking water containing veterinary products. See Chapter 6 for further details on the exposure of terrestrial environments to veterinary medicines.

7.2.2 ADDITIONAL SAFETY DATA AVAILABLE IN THE DOSSIER

As mentioned in earlier chapters, the potential adverse effects of a medicine in terrestrial and aquatic environments should not be evaluated in isolation. The data package used to assess the efficacy and safety of a veterinary medicine under development is extensive. Safety data packages for medicines intended for livestock include the results of studies to test the safety of the medicine in the target animal species, which are typically cattle, pigs, and poultry. Toxicity data are used to evaluate the safety to the consumer of ingestion of animal tissues (e.g., muscle, kidney, liver, or milk) containing medicine residues (human food safety). Furthermore, an evaluation is conducted to determine the potential impact of veterinary medicine residues on the normal gastrointestinal tract flora of humans (microbial safety). Finally, data from toxicity studies are used to address whether the farmer should be concerned for his or her safety when the medicine is administered to the target animal species (user safety). All of these data should be considered in the ecotoxicity risk assessment. For example, target animal safety data of a product for broiler chickens may identify a very low risk of avian toxicity and, therefore, reduce concerns that product residues might adversely affect nontarget bird species (e.g., raptors or vultures) due to secondary poisoning. In short, a dossier or application contains a wealth of safety information beyond that provided for the ecotoxicity assessment, which should be borne in mind when predicting the potential for veterinary medicine residues to affect the environment negatively.

7.2.3 RESIDUE DATA AND DETOXIFICATION BY THE TARGET ANIMAL SPECIES

The metabolism of medicines in treated animals can occur via many routes. Mammalian species have a broad range of P-450 enzymes with the capacity to modify xenobiotics that may enter their bodies. Veterinary medicines are examples of intentionally introduced xenobiotics for which much is known about their metabolism in the target species. It is mandated by certain regulatory authorities that companies sponsoring veterinary products have sufficient knowledge of the metabolism of the medicines in the target species to set recommendations for acceptable daily intakes (ADIs) and maximum residue limits (MRLs) to ensure the safety to humans of ingested tissues containing veterinary medicine residues.

7.3   PROTECTION GOALS

Tests on medicine residues provide the data for the risk assessment process. These data are then used to develop risk management and mitigation procedures to protect the functionality and structure (e.g., species diversity) of the terrestrial ecosystem. For logistical reasons we propose that these protection goals are generally limited to the rhizosphere. The rhizosphere is that portion of the soil associated with plant roots and is, therefore, the site of many key interactions between soil microorganisms and plant species. This limitation seems reasonable, particularly given the importance of the rhizosphere to crop production. These protection goals for veterinary medicines are similar to those previously defined for other classes of chemicals, such as agrochemicals and industrial chemicals. However, the proposals presented for the rhizosphere also reflect the use of the products and routes of entry into the environment.

Degree of probable exposure needs to be considered when setting protection goals. Species subject to exposure may be on-site, off-site, or migratory. On-site species are confined to the area where inputs of veterinary medicines are expected, for example soil microbes, some arthropod species, and earthworms (although some migration of these latter two groups may occur at field edges). Off-site species are located out of the main area of exposure, but may provide source populations for reinvasion and recovery of the more intensively managed on-site areas where a significant level of impact may be observed, for example some of the more mobile arthropod species or small wild mammals. Migratory species are mobile and can be expected to leave and reenter the treated area. Such species may include birds, mammals, and flying insect species.

The nature of land use should also be considered when setting protection goals. Acceptable levels of impact may vary for lands managed primarily for food production versus lands managed to protect natural ecosystems. With this consideration, we provide suggestions for experimental studies in Table 7.1 that are consistent with recommendations in the VICH phase II tier A risk assessment guidance document.

Four categories of land use are identified for illustrative purposes:

1) *Arable lands.* These lands are intensively managed for crop or forage production. Vegetation will be monocultures of nonnative species subject to very high levels of soil disturbance. Inputs usually are frequent and may include agrochemicals (e.g., herbicides, insecticides, and fungicides), fertilizers, and irrigation. The protection goal is to preserve the functionality and integrity of these lands for crop production. There is little consideration for the conservation of native species. Agronomic practice (e.g., deep ploughing and removal of hedgerows to increase field size) will have a significant impact on flora and fauna (e.g., earthworm populations are significantly depleted in arable lands subject to regular

ploughing). Contamination by veterinary medicines primarily occurs when manure or slurry from treated animals is removed from confinement facilities (e.g., dairies, cattle feedlots, or piggeries) and applied as fertilizer to these lands.

2) *Pastures for livestock production.* These lands include pastures managed primarily to produce food animals (e.g., beef cattle) or their products (e.g., milk). Such pastures frequently are sown with nonnative species of plants. There is a lower level of soil disturbance than that in arable lands, although inputs may still include agrochemicals and fertilizer. There is a greater opportunity to protect native species in these systems, although this is not the main objective. Contamination is most likely to occur when slurry from treated animals is applied as fertilizer or when dung is directly deposited by treated animals grazing these pastures.

3) *Pastures for livestock production and conservation of native species.* These lands are pastures managed jointly for both livestock production and the conservation of native species and natural ecosystems. There are little or no inputs. Examples include organic farms or lands held by the UK National Trust. Contamination is likely to occur only via the deposition of dung from treated livestock grazing on these lands.

4) *Natural protected systems.* These lands are managed primarily to protect species diversity and the functionality of natural ecosystems. Grazing by livestock is permitted only if there is no adverse effect on the primary objective. Examples include moorland, designated wilderness areas or sites of special scientific interest (SSSI), and national parks. Contamination is expected only via the deposition of dung from treated livestock grazing these lands. There is no active management of the grazing beyond the introduction and relocation of the animals.

We suggest that veterinary products could be labeled voluntarily to indicate their "environmental profile." Positive profiles would identify, for example, products with a very short half-life in soil and a low toxicity to arthropod species. Such products would be better suited for use in systems managed to protect natural ecosystem function (categories 3 and 4, above). Products with negative profiles would be more suited for use on arable lands or pastures for livestock production (categories 1 and 2).

Note that the four land categories identified in Table 7.1 are used to illustrate contrasting situations for which different priorities may be given to protect a system's function versus its natural diversity. In reality, there will not be distinct categories but rather a gradient across the full range. This conceptual model is intended to provide an additional tool to categorize the level of risk acceptable under different classes of land use compatible with existing legislation (e.g., US endangered species legislation, the Canadian Environmental Protection Act, and the EU Habitats Directive).

TABLE 7.1

Changing emphasis of protection goals across a gradient of land use: illustrated with four categories

Arable lands	Pastures for livestock production	Pastures for livestock production and conservation of native species	Natural protected systems
Functionality	Revise protection goal emphasis →	→	Structure
Phytotoxicity — crop species	Phytotoxicity — monocot and forage species	Phytotoxicity — monocot and dicot species	Phytotoxicity — monocot and dicot species
Earthworms	Earthworms	Earthworms	Earthworms
Soil arthropods — collembola and soil mites	Soil arthropods, for example collembola, soil mites, Aleochara, and dung fauna (fly and beetle)	Soil arthropods, for example collembola, soil mites, Aleochara, and dung fauna (fly and beetle)	Soil arthropods, for example collembola, soil mites, Aleochara, dung fauna (fly and beetle), and site-specific species
Soil microflora C/N cycling	Soil microflora C/N cycling	Soil microflora C/N cycling	Soil microflora C/N cycling
Data evaluation	Data evaluation	Data evaluation	Data evaluation
Apply VICH scenario for intensively reared animals	Apply VICH scenario for intensively reared animals and/or pasture depending on product type	Apply VICH scenario for pasture animals	Apply VICH scenario for pasture animals
Regional/country scale	←	→	Site specific

7.4   TIERED TESTING STRATEGY

The proposed testing strategy identified in Table 7.1 reflects current recommendations in VICH phase II tier A. Toxicity is evaluated in four major taxonomic groups that comprise plants, earthworms, nontarget arthropods, and soil microflora. Evaluation of the latter is achieved using a nitrogen transformation study. However, several modifications of the VICH protocol are proposed. Selection of test plant species should reflect land use. Crop species should be considered for arable lands. In contrast, native or noncrop species could be considered for assessments on pastures or natural protected systems. Soil arthropods of particular interest in arable lands would include collembolans and soil mites. Arthropods of interest in pastures also include species associated with livestock manure, for example dung beetles, coprophilous flies, and *Aleochara* spp. (rove beetles). Additional site-specific species may warrant special investigation in natural protected systems.

The VICH guidance recommends higher tiers of testing when the data evaluation indicates an unacceptable level of risk. However, the guidance document does not fully describe how these tests are to be conducted or how the endpoints are to be monitored. Generic study designs for tiers A, B, and C are proposed and compared in Table 7.2.

7.5   JUSTIFICATION FOR EXISTING TESTING METHODS

The justification for use of the testing methods (OECD and ISO) included in phase II must be understood in the context of the VICH negotiation process. It is accepted that other standardized methods (e.g., those of the American Society for Testing and Materials [ASTM], British Standards Institution [BSI], Office of Prevention, Pesticides and Toxic Substances [OPPTS], and USEPA) exist that may be appropriate to assess the potential impact of veterinary medicine residues on nontarget species in the terrestrial environment. Some of these other testing protocols are described later in this chapter. VICH adopted these specific study protocols because the OECD and ISO are internationally recognized bodies that periodically review and update their test protocols. In addition, some regions that were a party to VICH were unable to accept tests other than final OECD protocols or ISO studies. Notwithstanding this, the studies included in phase II should provide data sufficient in most cases to assess the potential impacts of veterinary medicine residues on nontarget species.

7.6   USE OF INDICATOR SPECIES

The concept of "indicator species" is well established for standard regulatory testing. The standard guidelines (OECD, ISO, etc.) have been developed and validated for representative indicator species for both aquatic and terrestrial species. The selection of the recommended species has been based on a number of considerations, including the following:

TABLE 7.2
Generic study designs for tiers A to C

	Tier A	Tier B[a]	Tier C
Objective	Basic toxicity evaluation Core data set	Higher tier effects evaluation	Usually field-based/site-specific effects evaluation
Study design	• Standard OECD or ISO methods • Dose response • Test compound spiked directly into soil or dung • Usually conducted using the technical active ingredient	• Based on standard guideline methods • Evaluating impact of dung residues in modified test systems under laboratory conditions • Selected doses based on PECs or natural dung residue levels • Test compound introduced into the test system in the form of residues from treated animals • Study conducted using proposed formulated product or API • Can be used to assess duration of effects using dung from treated animals over a period of time • Additional species to generate SSD	• Study-specific protocols, designed to address the issues of concern • Evaluate appropriate endpoints with reference to proposed product use • Studies usually conducted under field conditions, for example dung beetle function — degradation of cow pats, soil function, litter bag studies, and arthropod diversity impact
Endpoints	LC/EC_{50} values/NOEC These endpoints are used in the derivation of the PNEC and fed into the risk assessment.	NOEC This endpoint is used in the derivation of the PNEC and fed into the risk assessment.	Ecological function/biodiversity evaluations (endpoints defined depending on the issues of concern from previous levels of testing)
Data use	Tier I risk assessment	Refined risk assessment, reduced safety factor	Refined risk evaluation, reduced safety factor
Options	Go on to higher testing or accept risk mitigation/labeling limitation.	May confirm no effects under more realistic conditions of exposure or may indicate possible duration of adverse effects that may then be incorporated into an appropriate risk management strategy or labeling.	May confirm no effects under more field use conditions of exposure or may indicate possible duration of adverse effects that may then be incorporated into an appropriate risk management strategy or labeling.

a Tier B in the VICH phase II guidance document for plants is defined as 2 additional species and, for the soil nitrogen transformation, extension of the tier A study to 100 days. The testing of residues in dung from treated animals could be considered an optional extra tier B following the proposed refinements, if required.

- *Availability.* Can the organisms be cultured in the laboratory or be obtained from commercial suppliers?
- *Amenability.* How easy are the organisms to handle and maintain under laboratory conditions?
- *Appropriateness.* Is the species relevant for the part of the environment it is being used to represent, are there appropriate endpoints to monitor, and is it relatively sensitive to toxicants in a reproducible manner?

It is generally accepted in the area of environmental testing and effects evaluations that only a relatively limited number of species can be tested to represent the wider environment. To address this, the data from these standard tests are then subject to the application of additional assessment or safety factors to derive predicted no-effect concentrations (PNECs) to allow for potential species variability.

In addition to this interpretation of "indicator species" as those used for standard laboratory studies, the term can also be applied to species used as bioindicators in the field situation, which is relevant to higher tier field-based monitoring studies. Evaluating soil quality by measuring soil organisms has gained broad scientific acceptance. The presence or absence of indicator species, for example, may be a useful tool in evaluating the effects of veterinary medicines. The use of bioindicator species is being considered as an alternative extrapolation tool to whole ecosystem monitoring (Muys and Granval 1997).

Indicator species should provide information about the environment that is not readily apparent or is too costly to obtain in other ways. There may be at least two basic types of "species indicator" applications. The presence of particular rare species can be used to indicate the co-occurrence of other rare species that are not inventoried directly. Alternatively, the local species richness of one group of taxa can be used to represent the local species richness of the total taxa. Whereas the first approach may be used to delineate potential nature reserves, the second approach is more likely to be used to understand the pattern of biodiversity across the landscape.

The Nematode Maturity Index (NMI) is a widely used example of an indicator (Bongers 1990; Yeates 1994), although it has not yet been adopted in many nationwide monitoring programs. Calculation of the NMI is based on the proportion of nematodes with different levels of tolerance for disturbance. Low NMI values are often found in soils subjected to intensive agricultural production methods. Mid-range NMI values suggest a more diverse soil community and often reflect such practices as crop mixtures and rotations and no-till farming, whereas high NMI values are rarely found on cultivated lands.

Approaches using indicator species should frequently monitor selected groups of species representing different trophic levels for changes in population size and structure. Such changes could identify more pervasive effects on the larger set of species in the ecosystem. However, the implicit assumption that the observed changes are linked to veterinary medicine use is not directly tested in such an approach. It should therefore be considered in association with other data (e.g., toxicological data) to explain the observed changes.

7.7   SHORT-TERM AND SUBLETHAL EFFECTS TESTS

Tier A laboratory-based toxicity studies generally represent a worst-case scenario with enforced exposure to the compound under test. However, short-term bioassays, which are usually performed during only part of the entire life span of the test organism, may underestimate the adverse effects of exposure. Adult insects exposed to sublethal concentrations of a toxicant may exhibit loss of water balance, disrupted feeding and reduced fat accumulation, delayed ovarian development, decreased fecundity, and impaired mating (Floate et al. 2005). However, immature insects generally are more susceptible than adults and may exhibit additional effects of toxicant exposure including reduced growth rates, physical abnormalities, impaired pupariation or emergence, or delayed development (Floate et al. 2005). Ivermectin residues at levels that only marginally affect the survival of the dung beetle *Euoniticellus intermedius* can delay juvenile development by 7 weeks (Kruger and Scholtz 1997). Delays of this magnitude may result in adult emergence at a time of the year when conditions are less conducive to development or survival. In addition, sublethal effects of toxicant exposure experienced by individuals of the current generation may be expressed in subsequent generations via reductions in the fertility or size of females in the subsequent generation (Kruger and Scholtz 1995; Sommer et al. 2001). Toxicity studies combining chronic exposure of adult individuals with exposure of the more vulnerable offspring are therefore more likely to capture potential effects at the population level.

Long-term or chronic exposure to medicines and assessments of sublethal effects are often needed to elucidate fully the potential risk of substances that do not rapidly disappear from the soil.

7.8   TIER A TESTING

The design of terrestrial ecotoxicity studies should take into account the following information on the parent compound: physicochemical properties, fate, metabolism and excretion data, and the analytical methods for detection of the parent compound. Variations between regional regulatory authorities that should also be considered include the treatment regime (e.g., number and frequency per year, dosage, and route of administration) and environmental factors (e.g., climate and soil type). These considerations are also important for the interpretation of the test results, and appropriate studies are discussed in detail in OECD guidelines and in Chapter 6 of this book. The basic considerations for experimental design and interpretation are briefly discussed below.

7.8.1   Physicochemical Properties

Studies to determine solubility in water, dissociation constants in water (pK_a), the UV-visible adsorption spectrum, and the n-octanol/water partition coefficient (K_{ow}) for the parent compound are required in tier A of VICH. As well as being important data for use in the derivation of predicted environmental concentration

(PEC) values through modeling (see Chapter 6), they also provide valuable information that can be utilized to decide the appropriate design of the laboratory-based fate and effects studies (e.g., selection of solvents for spiking and selection of concentrations for aquatic-based studies).

The potential for bioaccumulation is based on the K_{ow} value and molecular weight. This information may also be used to evaluate the potential for secondary poisoning.

7.8.2 FATE

Studies to determine soil adsorption and desorption (coefficients K_d and K_{oc}) and soil biodegradation are recommended under tier A in the VICH phase II guidance document. Hydrolysis and photolysis studies are optional.

Interpretation of results from terrestrial effects studies requires knowledge of the bioavailability of the test substance. Many veterinary medicines are compounds with pH-dependent dissociable groups, and thus, under conditions where the test substance is a charged species, adsorption to soil may be affected. The pK_a, K_{oc}, and K_d values are used to determine the potential for binding to soil.

Data from metabolic and excretion studies on target species are used in conjunction with biodegradation studies to determine the PEC values in soil and dung (see Chapter 6). These studies can also be used to assess the need for and design of studies on metabolites and degradation compounds. The PEC values can be used to assist in the identification of appropriate test concentration ranges, particularly in higher tier studies.

The tier A effects studies are primarily standard OECD or ISO guideline methods, which are dose–response, laboratory-based experimental systems. The value of data derived at this level of testing is that the test conditions are well defined, which allows for a reproducible study design. This means that data generated using different test compounds can be compared to give a toxicity ranking. However, these studies were originally designed for evaluation of the toxicity of industrial and agrochemical products. It can be argued, therefore, that they do not always offer the most appropriate route of exposure for veterinary medicines. The following sections provide some background to the standard guideline studies and recommended test species.

7.8.3 MICROORGANISMS

Tests on specific microorganisms (e.g., pure culture maximum inhibition concentration tests) or functions carried out by microbial species are used as surrogates to assess the potential effects of veterinary medicine residues on processes mediated by these organisms (e.g., biogeochemical cycles). These cycles are important not only in pristine, natural environments but also in terrestrial environments used for intensive food production (Table 7.1). In VICH phase II, the recommended test is OECD 216. This test assesses the potential impact of veterinary medicine residues on the microbially mediated process of nitrogen mineralization. The rationale for preferring this test versus a test on potential impacts on carbon mineralization (e.g., OECD 217) is that fewer microbial species in soil catalyze the conversion of organic nitrogen to nitrite and nitrate as opposed to those capable of converting organic compounds (e.g., glucose) to inorganic products through the process of mineralization. It is generally recognized that tests to assess the impact of compounds such as veterinary medicine residues on microbial function are preferred over tests on individual species, given that the latter may not be truly representative of endogenous species.

7.8.4 PLANTS

Tests on individual plant species are used as surrogates to evaluate the potential effects of veterinary medicine residues on plant species important in different terrestrial environments, such as those mentioned in Table 7.1. The OECD 208 study is recommended in VICH phase II for this assessment. The number of species selected and the category (1 of 3) from which they are drawn are most often determined based on convenience and regulatory considerations, rather than the relevance of the test species to the actual species present in specific terrestrial habitats. It is suggested, therefore, that some consideration of the type of terrestrial habitat of interest to the assessor (Table 7.1) helps determine the choice and number of species selected for inclusion in a given OECD 208 study.

7.8.5 EARTHWORMS

Earthworms (order Oligochaeta) are routinely used in soil ecotoxicology evaluations. About 1800 species occur in 5 families with global distribution of the order. Earthworms most common in North America, Europe, and Western Asia belong to family Lumbricidae, which has about 220 species.

Earthworms mainly derive their nutrition from organic matter in a wide variety of forms that may include plant material, protozoans, rotifers, nematodes, bacteria, fungi, and decomposed material (Curry 1998). The feeding, burrowing, and cast-forming characteristics of (particularly) endogeic and anecic worms thoroughly mix organic and mineral components of the soil (Edwards and Shipitalo 1998), and increase its porosity and permeability. The extent to which soil porosity is affected depends largely on the number of earthworms in the soil, their spatial distribution, and their size. Increased porosity reduces soil erosion and can increase water percolation through the soil profile.

The inception, ring testing, and standardization of the acute earthworm toxicity test (OECD 207) within the OECD regime have since 1984 comprised a catalyst for the emergence of earthworms as 1 of the key organisms in environmental toxicology (Spurgeon et al. 2004). It was followed 20 years later (2004) by a chronic toxicity test focusing on sublethal reproductive effects (OECD 222). The commonly used test species *Eisenia fetida*, *Eisenia andrei*, and *Eisenia veneta* belong to the class of manure worms and red worms. They can adapt to living in many different environments and will eat almost any organic matter at some stage of decomposition. These worms can be found in manure piles or in soils containing large quantities of organic matter and are also bred commercially.

7.8.6 COLLEMBOLANS

Collembolans or "springtails" are small wingless insects with global distribution that occur on and below the soil surface. They are the most abundant group of insects. A square meter of temperate grassland may contain at least 50000 or even up to 200000 individuals comprising 20 to 30 different species. Their diets typically consist of fungal hyphae and organic detritus such that they play an important role in the decomposition of organic material and recycling of nutrients (Filser 2002). The presence of springtails is therefore important for maintaining a well-functioning agricultural soil. Furthermore, their widespread distribution and large diversity in most ecosystems make them suitable surrogates for evaluating potential changes in biodiversity.

The chronic toxicity test with the soil-dwelling collembolans *Folsomia candida* and *Folsomia fimetaria* was developed during the early 1990s (Krogh et al. 1998; Løkke and van Gestel 1998; Wiles and Krogh 1998) and was adopted as an ISO standard in 1999 (ISO 11267). At the time of going to press, the draft OECD guideline is undergoing review and commenting.

7.8.7 DUNG FAUNA

Descriptions of insects in cattle dung and the potential adverse effects of veterinary medicines are presented in more detail elsewhere (e.g., Strong 1993; Wratten 1996; Floate et al. 2005; Floate 2006). In brief, dung pats support numerous and diverse species of insects, mites, nematodes, earthworms, fungi, and microorganisms. The majority of these species are either innocuous or beneficial by virtue of accelerating the process of dung degradation. Only a few taxa are nuisance or pest species.

Fresh dung is colonized in a series of successional waves. The first wave is composed primarily of adult flies. They arrive within minutes with peak visitation, usually within the first few hours of pat deposition. Eggs laid by these flies produce a new generation of adult flies in 10 to 20 days. The second wave is represented primarily by adult dung-feeding beetles (e.g., Scarabaeidae), whose numbers peak usually during the first week of pat deposition. Egg-to-adult development time of beetles may take weeks to months. Flies and beetles visiting the pat often carry phoretic nematodes and mites, whose numbers begin to increase about 10 days after arrival at the dung and continue to grow for several weeks. The first and second waves of succession coincide with the arrival of wasps parasitic on immature flies and of beetles predaceous on the immature stages of previous colonizers. Fungivorous beetles colonize pats at later stages of decomposition to feed on fungal hyphae and spores. Coprophilous insects are unlikely to colonize dung beyond 45 days in temperate pastures or beyond 14 days under many tropical conditions. The final colonization phase occurs with the breakdown of the interface between the dung and the soil surface. This process provides access into the dung of soil-dwelling organisms (e.g., earthworms and bacteria) to complete the breakdown of the dung to its component parts.

Variation in biotic and abiotic factors, plus differences in animal management practices, affects the extrapolation of observations across geographic regions. With reference to dung beetles (Scarabaeidae), regions may differ in species composition and the dominance of functional groups. Functional groups include "dwellers," "tunnellers," and "rollers." Degradation of dung pats by dwellers typically occurs via larval feeding during a period of weeks to months. Degradation by tunnellers and rollers is through the actions of adult beetles, with complete pat breakdown and dissipation possible within hours or days. In regions dominated by tunnellers and rollers, delays in breakdown and dissipation associated with the use of veterinary medicines may be apparent in a matter of days. In regions dominated by dwellers, such effects may not be apparent for weeks. Hanski and Cambefort (1991) provide an excellent overview of dung beetle ecology with comparisons of dung beetle communities between geographic regions worldwide. With reference to earthworms, high numbers are common on European pastures, where they can be the main agency of dung removal. Conversely, earthworms are largely absent from large regions of North America, such that insects often are the main agents of dung pat degradation.

Species composition and biotic activity in dung are strongly affected by season. Insect and earthworm activity tends to be highest when conditions are warm or wet. Many species of dung-dwelling beetles exhibit a single peak of adult activity in the spring corresponding to the emergence of overwintered individuals. Dung pats deposited on pasture during this time usually are most rapidly degraded. Other species exhibit peaks of both spring and autumn adult activity, with the latter corresponding to the emergence of the new adults developed from eggs laid in the spring. Flies typically have several generations per year, with peak numbers occurring in late summer before the onset of cooler or drier conditions. Recognition of seasonal variation may be required to optimize the design of tier C tests to assess the effects of fecal residues on dung community structure and function under field conditions.

The effects of veterinary medicine residues in the dung of treated livestock should be considered not just within a broader framework of regional and seasonal variation in dung organism composition and activity but also with regard to abiotic factors and animal management practices (Figure 7.1). Such consideration increases appreciation of the complex interactions affecting dung pat degradation and the difficulty in extrapolating effects across broad geographic regions. The intent for which pastures are managed (e.g., livestock production versus protection of native biodiversity) affects stocking rates. Stocking rates affect the density of dung pats and the likelihood of these pats being disrupted by trampling. Forage type (native vegetation versus tame grasses) affects dung moisture content, which affects the size and shape of the pat upon deposition. Location of deposition (woodland versus grassland) can affect pat degradation directly, by influencing the rate of dung desiccation, and indirectly, by influencing the composition and number of insect colonists.

Tier A acute toxicity studies for representative species of dung flies and dung beetles are currently under development through the Dung Organism Toxicity

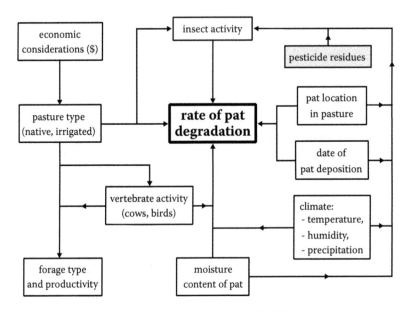

FIGURE 7.1   Abiotic and biotic factors that affect the degradation of cattle dung pats on pasture. *Note*: Regional variation in some of these factors (e.g., pasture type, weather, and type of dung beetle species) can confound detection of delayed degradation that may be associated with use of veterinary medicines. *Source*: Modified from Merritt and Anderson (1977).

Testing Standardization (DOTTS) Group, which is operating under the auspices of SETAC. The tier A studies are being designed to monitor survival of the test species in standardized laboratory test systems, utilizing spiked dung in a dose–response-style study. The validated test method will be issued as an OECD guideline in the case of the dung fly test, and as an OECD guidance document in the case of the dung beetles. At the time of going to press, the draft OECD dung fly guideline has been approved for publication. The draft dung beetle guidance document is currently under review by OECD members.

7.9   TIER B TESTING

It is proposed that the experimental studies conducted at tier B broadly follow the tier A methods, but with revision of the route of introduction of the test compound. For veterinary medicines it is proposed that the route of introduction, where appropriate, should reflect the natural route of introduction into the environment. Many products will enter the environment via the animal in the feces. Therefore, it is suggested that tests on earthworms, collembolans, and dung fauna use dung from animals treated with the product in accordance with the label recommendations. The residue-contaminated dung could be incorporated at rates equivalent to the modeled PEC, assuming addition to dung incorporation from treated animals at pasture or from slurry spreading as appropriate. It may also be useful in such studies to incorporate dung from animals treated at rates higher than recommended, but within the range of animal safety identified in target animal safety studies. Results from such studies would enable some evaluation of the safety factor from higher application rates.

Dung from treated animals also could be collected over an appropriate time course to evaluate the potential duration of residual excretion. This information could be used to set label recommendations regarding the period of animal housing posttreatment. It would also indicate the time post treatment that residues in fresh dung deposited on pastures may affect biotic activity.

For intensively reared animals it may be more appropriate to adopt this higher tier testing strategy with slurry from treated animals. This approach could be used to determine the optimum period for slurry storage prior to spreading to minimize the impact on nontarget species.

7.10   TIER C TESTING

Under the VICH guidelines, tier C studies are recommended when the risk quotient values exceed 1 after tier B ecotoxicity testing and recalculation of the PEC. VICH does not provide guidance for the types of studies that are suggested, and so applicants must seek guidance from the regulators in the region where registration is being sought. Often the concerns raised at tier C are specific to a region or are site specific within the region. The types of studies required to address these concerns are outside the scope of approved guidelines such as those of the OECD and ISO. Discussions between the applicant and the regulatory authority are essential for considering the design of such studies. Application of sound scientific principles is the foundation for the design and conduct of these studies and must be one of the key criteria for the acceptability of the results of these studies.

Tier C studies may be required to address concerns regarding specific species that are considered to be sensitive for ecological (e.g., endangered species) and agricultural (e.g., functional species) consideration. Special concerns may be raised for protected natural systems (e.g., National Trust land in the United Kingdom). Surrogate test species may have to be used in laboratory studies if the species of interest cannot be easily cultured or maintained under laboratory conditions. Some of the options for testing at tier C are considered here.

7.10.1   Mesocosm and Field Testing

Terrestrial mesocosm and field studies can be used to examine the long-term effects of veterinary medicines under conditions of treatment, although these types of studies are generally rarely required for veterinary medicines (see Chapter 5 for a discussion of aquatic microcosms and mesocosms). They can be used to examine multiple species effects. The advantages and disadvantages of each test system must be considered before designing a study to address concerns that remain after a Tier B assessment.

Mesocosms are semiartificial systems with a limited sampling area. However, the level at which conditions can be controlled (e.g., application and soil moisture) enhances reproducibility. Radiolabeled chemicals can be used in these systems. The use of several mesocosms increases the number of "test" sites.

Field studies are more reflective of actual use conditions. However, variability at sampling points may result from local differences in soil conditions over a larger test area and uneven distribution of the medicines during manure application. Variation in rainfall from seasonal averages may affect interpretation of the results. Consideration should be given to the use of an irrigation system, as used in some pesticide field studies, to ensure a minimum level of precipitation representative of average rainfall levels during the period of the study. Knowledge of the history of previous chemical use (e.g., veterinary medicines or pesticides) at test sites is essential to ensure that observed treatment effects can be attributed solely to the test material. Persons not associated with the study should be restricted from the study area. Cost may limit the use of field studies, which are usually more expensive than mesocosm studies.

7.10.2　Testing of Additional Species

Concerns regarding sensitive species, including taxa that may be listed as endangered, are regional and often site specific. Some regulatory systems require that consideration of endangered species be taken into account in the risk assessment. Although direct testing of these species is usually not feasible, this question may be addressed through the use of a wider range of species or an indigenous surrogate species. Tests on additional species can help to define the species sensitivity distribution (SSD; see Chapter 5), which can then be used in the risk assessment to help reduce safety factors. (See Section 7.11 on calculation of PNEC concentrations.)

7.10.3　Monitoring Studies

Monitoring studies should be considered if concerns remain after laboratory, mesocosm, or field studies, or if these studies are considered inappropriate to address concerns over a specific risk. Monitoring generally should be limited to "on-site" species (excluding birds) for which a risk has been identified. Studies of impacts on bird species should examine nesting populations in the immediate vicinity of treated sites, with the monitoring area defined by the feeding range for the species being studied. As noted for field studies, it is necessary to have a historical baseline for population levels for the sites monitored and the history of treatment with veterinary medicines or pesticides. Similarly, there should be restricted access to the monitored areas. Monitoring studies should be multiyear to account for yearly variation in climatic factors (e.g., temperature and precipitation) and population trends.

It is anticipated that tier C studies will usually be conducted under field-type conditions. There should be no prescriptive methods for these studies, but protocols should be developed on an individual basis to address the specific issues of concern. Examples may include the following:

- *Litter bag studies to evaluate soil function.* Such studies monitor the degradation of organic straw parcels over a period of up to 1 year. For a veterinary medicine the existing guidance for agrochemical products could be modified to reflect the route of compound introduction via manure or slurry. The degradation of the straw in these studies reflects the total function of the system and degree of biodiversity.
- *Long-term studies to monitor the effect of veterinary medicines on earthworm populations.* Such studies could be performed using a modified version of the ISO standard method with the test compound applied to soil in the slurry or manure as appropriate.
- *Studies to monitor effects of veterinary medicines on the number, diversity, and activity of arthropods in dung deposited by treated animals on pasture.* These studies also could monitor effects of residues on rates of dung degradation.

Tier C tests sacrifice sensitivity to improve reality. Background "noise" and system variability will likely confound detection of partial effects. Hence, use of specialized statistical methods (e.g., principal response curves, or PRCs) to evaluate population trends may be necessary. In many cases, it may be more appropriate for tier C tests to target the functionality of the system and community impact, rather than effects on individual species.

7.11 CALCULATION OF PNEC CONCENTRATIONS AND USE OF ASSESSMENT FACTORS

Within the framework of risk assessment, toxicity data can be used to calculate a predicted no-effect concentration. This is compared with the predicted environmental concentration to establish the risk quotient, which forms the basis for most regulatory decision making (see Chapter 3). PNEC values are derived using assessment factors (e.g., by applying a factor between 10 and 1000 to the endpoint of each test). The assessment factors (AF) are fixed for tiers A and B of the VICH guidelines. No specific AFs have been assigned to higher tier testing. Taking into account the increased realism and added information from higher tier tests, AFs between 1 and 10 are reasonable and consistent with the approach applied in other regulatory situations (e.g., for pesticides).

Probabilistic approaches (e.g., species sensitivity distributions, or SSDs) have been suggested in recent years to derive PNEC values (see Chapter 5 for more details). Dossiers submitted for an authorization of a veterinary medicine are unlikely to include sufficient terrestrial toxicity data to calculate SSDs. However, for existing products, SSDs may be a useful tool for the terrestrial environment if sufficient data can be obtained from the open literature on soil-dwelling organisms and processes. A PNEC is derived from an SSD by estimating the maximum concentration that potentially affects a predefined fraction of the species. This fraction of potentially affected species is typically defined as 5%, also referred to as the HC5 (Aldenberg and Jaworska 2000; Forbes and Callow 2002).

It may be useful to shift the focus from estimation of the concentration that affects a predefined fraction of all species (e.g., HC5) to estimation of the potentially affected fraction (PAF) of species by predicted soil concentrations. Whereas a comparison of PEC and PNEC provides the basis for deciding if risks are acceptable, the PAF estimation may provide better insight into the magnitude of a potential effect.

7.12   METABOLITE TESTING IN TIERS A AND B

Ecotoxicity studies required under the VICH guidelines use the parent compound as the test material. A total residue approach was adopted to account for the potential toxicity of metabolites of veterinary medicines. This approach assumes that metabolites are generally less toxic than the parent. Consequently, there are no requirements for metabolite testing.

However, consideration should be given to use of a screening process to determine if there is a potential for persistent major metabolites (i.e., greater than or equal to 10% of the initial parent) to have significant toxicity and to select criteria for further investigation. Therefore, there is a need for a simple and cost-effective screening process to determine the degree of this potential toxicity prior to conducting toxicity studies on metabolites. Similarly, degradation products resulting from environmental processes (e.g., biodegradation) should also be considered. These degradates may be formed in the dung, manure, or soil and, in some instances, may be similar or identical to excreted metabolites. Toxicity studies on similar metabolites could provide surrogate information, but such data are not typically available. The use of predictive models (e.g., QSARs) for ecotoxicity may be a useful tool for this purpose. However, most ecotoxicity models are based on aquatic organisms (e.g., fish, algae, and daphnids). Furthermore, use of models designed for industrial chemicals may be inappropriate for veterinary medicines, which may be ionic compounds or have unique chemical moieties that are not included in the validation of these models.

When concerns exist that a major metabolite or an environmental degradation product may be more toxic than the parent compound, the fate of the metabolite or degradation product should be considered by using modeling or by reexamining the data for the parent molecule. The estimated values for persistence, adsorption, solubility, and bioaccumulation potential of these substances should be examined before proceeding to ecotoxicity studies. Compounds with a short half-life and strong adsorption to soil should be excluded from further testing as there is unlikely to be significant exposure to these compounds for terrestrial species (see Chapter 6). For significant degradates produced in soil (> 10%), the ability to synthesize and radiolabel the compound comprises practical considerations that must be taken into account before further testing. If these are readily synthesized, the toxicity test outlined for the parent compound in tier A should be conducted. For metabolites, the use of treated manure can be considered appropriate for determining toxicity if a parent compound degrades relatively rapidly so that parent and metabolite effects are distinguishable. The treated manure should be stored

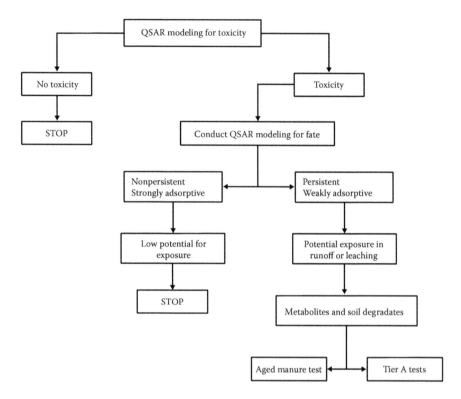

FIGURE 7.2   Screening schemes for testing metabolites and soil degradates.

for a period of 2 to 3 half-lives for the parent. This type of test can determine the toxicity of a metabolite if there is only 1 major metabolite. For parent compounds with more than 1 major metabolite, testing can determine the toxicity of all the major metabolites, but the test will not be able to determine the toxicity of an individual metabolite. A summary of the proposed testing strategy for metabolites is shown in Figure 7.2.

7.13   SECONDARY POISONING

Secondary poisoning may occur when a substance has the potential for toxicity and bioaccumulation in species that are consumed by birds and mammals. The potential for food chain effects should be examined for these compounds. The impact of all food sources, both terrestrial and aquatic species, should be considered. Calculations for secondary poisoning based on predicted values for daily consumption of prey and estimated bioaccumulation for different species of prey in the food chain are routinely employed for the evaluation of pesticides (European Commission 2002). Where the veterinary medicine is intended for treatment of an avian species, data on the toxicity of the parent compound conducted with that species may be used for assessing the potential for secondary poisoning.

7.14   BOUND RESIDUES

Organic pollutants and heavy metals may undergo aging processes that alter their mobility, degradation rates, toxicity, and uptake in biota. It is common in field studies to observe a relatively rapid disappearance of organic contaminants followed by a subsequent slow disappearance of the residual fraction — a so-called hockey stick-shaped degradation curve. The increasing pollutant retention over time, or "aging," usually occurs in soil and sediment and may significantly reduce bioavailability. Strong sorption and slow release processes are responsible for this sequestration of hydrophobic pollutants. Major processes involved are diffusion into nanopores and sorption (adsorption and partitioning) to soil organic matter. The magnitude and pace of sequestration depend on a number of parameters, with soil type (e.g., quantity and quality of organic matter), climatic conditions, and physicochemical properties of the contaminant being the most important. Time-dependent changes in sorption can be very important for the dissipation of several different classes of chemicals (Calderbank 1989; Führ 1987).

The sorption and desorption of substances are important for a number of processes in the soil system, including mobility, degradation, and uptake into biota. Sorption of organic substances depends on a number of parameters. However, the many functional groups associated with veterinary medicines make it more difficult to predict the behavior and fate of this group of substances compared to other organic substances.

There is general scientific consensus that inert and nonextractable residues are not ecotoxicologically relevant (e.g., Roberts 1984). However, it may be difficult to define and measure the fraction of substances of no concern. In some cases, it may not be residues of the parent compound that are bound to the soil, but rather the metabolites or the degradation products. If one is only relying on radioactivity from labeled parent compounds for information on contaminant concentrations, it will not be possible to distinguish between bound residues of the parent compound and residues of metabolites or degradation products, or even labeled carbon incorporated into microorganisms (see Chapter 6).

Bioavailability to soil organisms encompasses several distinctive phase transition and mass transfer processes (e.g., Lanno 2004; Jensen and Mesman 2006). One is the amount of substance potentially available for uptake. This is typically the fraction of chemical freely dissolved in the pore water and, to a certain extent, the fraction that easily desorbs from soil particles or dissolved organic material such as humic acid. This process is physicochemically driven and controlled by substance- and soil-specific parameters like log K_{ow}, pK_a, cation exchange capacity, pH, clay, and organic matter. Organism behavior, anatomy, feeding strategy, and habitat preference, together with physiologically driven uptake processes, can influence how much is actually taken up. For example, earthworms may take up lipophilic substances through ingested organic material. Differences in metabolism, detoxification, storage capacity, excretion, and energy resources also may have a large influence on how much of the substance is taken up and reaches the target of concern.

A number of physicochemical techniques have been developed to gain knowledge about the extent of pollutant retention and the fraction of contaminants available for biota, including microbial degradation. These include, for example, solid-phase micro extraction (Van Der Wal et al. 2004; Ter Lark et al. 2005), rapid persulfate oxidation (Cuypers et al. 2000), surfactant extraction (Volkering et al. 1998), cyclodextrin extraction (Reid et al. 2000; Cuypers et al. 2002), Tenax extraction (Cornelissen et al. 2001; ten Hulscher et al. 2003), semipermeable membrane devices (Macrae and Hall 1998), solvent extraction techniques (Chung and Alexander 1998; Tao et al. 2004), and the supercritical fluid extraction technique (Dean et al. 1995; Khan 1995). These techniques have primarily been developed and tested with organic pollutants like pesticides or PAHs. Very few data are available to support the use of any of these techniques in assessing the fraction of medicines available for uptake and toxic action in soil-dwelling species.

In conclusion, ecotoxicity studies inherently take bioavailability into consideration because biota only respond to the biologically active fraction of toxicants. However, including bound residues in the risk assessment of veterinary medicines would require evaluation of the bioavailability of the test compounds in the studies forming the basis of the PEC calculations (e.g., the biotransformation study, as discussed in Chapter 6).

7.15   ALTERNATIVE ENDPOINTS

Medicines are typically designed to have a specific mode of action. Hence, the efficiency of human medicines, in particular, potentially can be monitored by the use of substance-specific biomarkers. Biomarkers used in biology and ecotoxicology have been defined as "any biological responses to an environmental chemical at the individual level or below (cellular or molecular) or demonstrating a departure from the normal status" (Walker et al. 2001). Lysosomal membrane stability is a cellular marker for stress that has been widely used with various earthworm species (Svendsen and Weeks 1997; Scott-Fordsmand et al. 1998; Reinecke and Reinecke 1999; Spurgeon et al. 2000; Svendsen et al. 2004; Jensen et al. 2007). Biological indicators of adverse ecological changes could be changes in morphology, physiology, or behavior.

To be useful as risk assessment tools, biomarker or bioindicator responses should predict changes in the fitness of organisms and, by extension, the stability of their populations. One such biomarker may be fluctuating asymmetry (FA), which refers to small, random deviations between sides in an otherwise symmetrical organism. The level of these deviations has been reported to increase with the level of environmental stress (e.g., toxicants, temperature, and competition) encountered by an organism during its development. For larvae of the dung-breeding fly *Scatophaga stercoraria*, 50% failed to emerge as adults when reared in dung spiked with 0.001 ppm (wet weight) of ivermectin. However, enhanced levels of FA were detected in wing traits of flies reared with exposure to as little

as 0.0005 ppm ivermectin (Strong and James 1992). Its extreme sensitivity, potentially broad application, and relative ease of data collection make FA an attractive
tool for use as a bioassay. However, caution is urged regarding the use of FA in
the risk assessment procedure until a clear association has been demonstrated
between toxicant exposure and levels of FA. Other studies, including several on
insects breeding in dung with residues of veterinary parasiticides (Wardhaugh
et al. 1993; Floate and Fox 2000; Sommer et al. 2001), show no effect of toxicants
on FA. Furthermore, enhanced levels of FA may be biologically insignificant if
the fitness of exposed organisms is not otherwise affected (e.g., Floate and Fox
2000).

Biomarker or bioindicator systems potentially may be used as indicators of
risk when, for example, monitoring the effects of medicines in areas treated with
manure. However, many of the above responses are likely to be a general reaction
to stress rather than unique responses associated with exposure to the veterinary
medicine of interest. Potential confounding effects of other stress factors should
be taken into consideration with the use of appropriate control or reference populations when interpreting results.

There has been very little validation for the use of biomarkers in the risk
assessment of veterinary medicines. Additional validation may increase the use
of biomarkers in higher tier testing.

7.16 MODELING POPULATION AND ECOSYSTEM EFFECTS (E.G., BIOINDICATOR APPROACHES)

Large-scale, long-term, and multidisciplinary (e.g., involving entomologists, plant
ecologists, soil biologists, and economists) field studies ideally are required to
monitor the effects of fecal residues on populations of dung-dwelling insects and
the associated effects on dung degradation and pasture productivity. Such studies make the fewest assumptions but are hampered by logistical considerations.
Comparisons are required between a region in which all livestock are similarly
handled and treated versus an adjacent, equivalent region in which no livestock
are treated. Replication of such pairwise comparisons is required in different
geographic locations to extend the generality of findings under varying conditions including weather, insect fauna, and land use. Furthermore, such pairwise
replications likely will be required for an undefined number of consecutive years
to assess accumulated impacts of chronic exposure that may not be evident in the
first, second, or even third year of the study.

Ecotoxicological models can provide a practical and objective way to assess
the impact of veterinary medicines on the dung insect community at larger temporal and spatial scales. The models developed thus far (summarized in Cooper
et al. 2003; an example is described in Chapter 3) assess treatment impacts at
the scale of an individual farm, herd, or flock. These models demonstrate that
recommended use of at least some veterinary medicines can reduce populations
of dung-breeding species of insects within a given season, and they identify

key factors affecting the extent of these reductions. Key factors include product formulation (Wardhaugh et al. 2001), the proportion of livestock treated with the product (Sherratt et al. 1998; Vale and Grant 2002), and the degree of overlap between the period of fecal residue excretion and the seasonal activity of a given species (Wardhaugh et al. 1998, 2001; Vale and Grant 2002).

Estimating the likely within-season effects of veterinary medicine usage on dung fauna is challenging. Predicting the longer term impact of these effects on the average population sizes of a given species is even more problematical. Interactions among dung-dwelling species are complex. Furthermore, the demographic parameters that affect population size (e.g., fecundity and survival) are often density dependent (Sherratt et al. 1998). Therefore, although mathematical and computer models may have a future role in evaluating the impacts of particular parasiticide use patterns, they do not replace carefully conducted field studies.

7.17 RESEARCH NEEDS

We identified a number of topics requiring further investigation, but such investigations are likely to receive only limited attention in the absence of sufficient funding. For example, the development of tier A dung fauna toxicity testing methods has been in progress for some years under the auspices of the SETAC DOTTS group. The development of these methods is a high priority for the OECD. However, a lack of targeted funding has limited the number of laboratories participating in the validation of these methods. Within the participating laboratories, validating the methods is usually of secondary importance to work on funded projects.

Other areas requiring further study include the modeling of population and ecosystem effects, the validation of alternative endpoints (e.g., biomarkers), and the assessment of the biological relevance of bound residues.

REFERENCES

Aldenberg T, Jaworska JS. 2000. Uncetainty of the hazardous concentration and fraction affected for normal species sensitivity distribution. Ecotoxicol Environ Saf 46:1–18.

Bongers T. 1990. The maturity index: an ecological measure of environmental disturbance based on nematode species composition. Oecologia 83:14–19.

Calderbank A. 1989. The occurrence and significance of bound pesticide residues in soil. Rev Environ Contam Toxicol 108:71–101.

Chung N, Alexander M. 1998. Differences in sequestration and bioavailability of organic compounds aged in dissimilar soils. Environ Sci Technol 32:855–860.

Chung N, Alexander M. 2002. Effect of soil properties on bioavailability and extractability of phenarthrene and atrazine sequestered in soil. Chemosphere 48:109–115.

Cooper CS, Sherratt T, Boxall A. 2003. Modelling the impact of residues of ectoparasiticides and endoparasiticides in livestock dung on populations of dung flora and fauna: phase 1. Final Report to English Nature Project No. 31-004-12. Cranfield (UK): Cranfield University.

Cornelissen G, Rigterink H, Hulscher TEM, Vrind BA, van Noort PCM. 2001. A simple Tenax® extraction method to determine the availability of sediment-sorbed organic compounds. Env Tox and Chem 20:706–711.

Curry JP. 1998. Factors affecting earthworm abundance in soils. In: Edwards CA, editor. Earthworm ecology. Boca Raton (FL): CRC Press.

Cuypers C, Grotenhuis T, Nierop KGJ, Franco EM, de Jager A, Wulkens W. 2002. Amorphos and condensed organic matter domains: the effect of persulphate oxidation on the composition of soil/sediment organic matter. Chemosphere 48:919–931.

Cuypers C, Pancras T, Grotenhuis T, Rulkens W. 2002. The extraction of PAH bioavailability in contaminated sediments using hydroxypropyl-β-cyclodextrin and Triton X-100 extraction techniques. Chemosphere 46:1235–1245.

Dean JR, Barnabas IJ, Fowlis IA. 1995. Extraction of polycyclic aromatic hydrocarbons from highly contaminated soils: a comparison between soxhlet, microwave, and supercritical fluid techniques. Anal Proc 32:305–308.

Edwards WM, Shipitalo MJ. 1998. Consequences of earthworms in agricultural soils: aggregation and porosity. In: Edwards CA, editor. Earthworm ecology. Boca Raton (FL): CRC Press.

European Commission, Health and Consumer Protection Directorate-General. 2002. Guidance document on risk assessment for birds and mammals under Council Directive 91/4414/EEC. Working document. SANCO/4145/2000, final September 25. http://ec.europa.eu/food/plant/protection/evaluation/guidance/wrkdoc19_en.pdf.

Filser J. 2002. The role of Collembola in carbon and nitrogen cycling in soil. Pedobiol 46:234–245.

Floate KD. 2006. Endectocide use on cattle and faecal residues: an assessment of environmental effects in Canada. Can J Vet Res 70:1–10.

Floate KD, Fox AS. 2000. Flies under stress: a test of fluctuating asymmetry as a biomonitor of environmental quality. Ecolog Applic 10:1541–1555.

Floate KD, Wardhaugh KG, Boxall ABA, Sherratt TN. 2005. Faecal residues of veterinary pharmaceuticals: non-target effects in the pasture environment. Ann Rev Entomol 50:153–179.

Forbes VE, Calow P. 2002. Species sensitivity distributions re-visited: a critical appraisal. Hum Ecol Risk Assess 8:473–492.

Führ F. 1987. Nonextractable pesticide residues in soil. In: Greenhalgh R, Roberts TR, editors. Pesticide science and biotechnology. Oxford (UK): Blackwell Scientific, p 381–389.

Hanski I, Cambefort Y, editors. 1991. Dung beetle ecology. Princeton (NJ): Princeton University Press, 481 p.

Jensen J, Diao X, Scott-Fordsmand JJ. 2007. Sublethal toxicity of the antiparasitic abamectin on earthworms and the application of neutral red retention time as a biomarker. Chemosphere 58 (40):744–750.

Jensen J, Mesman M. 2006. Ecological risk assessment of contaminated land. RIVM report No. 711701047. ISBN No. 90-6960-138-9 978-90-6960-138-0.

Khan S. 1995. Supercritical fluid extraction of bound pesticide residues from soil and food commodities. J Agric Chem 43:1718–1723.

Krogh PH, Johansen K, Holmstrup M. 1998. Automatic counting of collembolans for laboratory experiments. Appl Soil Ecol 7:201–205.

Kruger K, Scholtz CH. 1995. The effect of ivermectin on the development and reproduction of the dung-breeding fly *Musca nevilli Kleynhans* (Diptera, Muscidae). Agric Ecosyst Environ 53:13–18.

Kruger K, Scholtz CH. 1997. Lethal and sublethal effects of ivermectin on the dung-breeding beetles *Euoniticellus intermedius* (Reiche) and *Onitis alexis Klug* (Coleoptera, Scarabaeidae). Agric Ecosyst Environ 61:123–131.

Lanno R, Wells J, Conder J, Bradham K, Basta N. 2004. The bioavailability of chemicals for earthworms. Ecotox and Envir Safety 57:39–47.

Løkke H, van Gestel CAM. 1998. Handbook of soil invertebrate toxicity tests. London (UK): John Wiley & Sons Ltd, 281 p.

MacRae JD, Hall KJ, 1998. Comparison of methods used to determine the availability of polycyclic aromatic hydrocarbons in marine sediment. Environ Sci Technol 32:3809–3815.

Merritt RW, Anderson JR. 1977. The effects of different pasture and rangeland ecosystems on the annual dynamics of insects in cattle droppings. Hilgardia 45:31–71.

Muys B, Granval PH. 1997. Earthworms as bio-indicators of forest site quality. Soil Biol Biochem 29:323–328.

Reid BJ, Jones KC, Semple KT. 2000. Bioavailability of persistent organic pollutants in soils and sediments — a perspective on mechanisms. Env Pollution 108:103–112.

Reinecke SA, Reinecke AJ. 1999. Lysosomal response of earthworm coelomocytes induced by long-term experimental exposure to heavy metals. Pedobiol 43:585–593.

Roberts TR. 1984. IUPAC reports on pesticides. 17. Non-extractable pesticide-residues in soils and plants. Pure Appl Chem 56:945–956.

Scott-Fordsmand JJ, Weeks JM, Hopkin SP. 1998. Toxicity of nickel to the earthworm and the applicability of the neutral red retention assay. Ecotoxicol 7:291–295.

Sherratt TN, Macdougall AD, Wratten SD, Forbes AB. 1998. Models to assist the evaluation of the impact of avermectins on dung insect. Ecol Model 110:165–173.

Sommer C, Vagn Jensen KM, Jespersen JB. 2001. Topical treatment of calves with synthetic pyrethroids: effects on the non-target dung fly *Neomyia cornicina* (Diptera: Muscidae). Bull Entomol Res 91:131–137.

Spurgeon DJ, Svendsen C, Rimmer VR, Hopkin SP, Weeks JM. 2000. Relative sensitivity of life-cycle and biomarker responses in four earthworm species exposed to zinc. Environ Toxicol Chem 19:1800–1808.

Spurgeon DJ, Weeks JM, Van Gestel CA. 2004. A summary of eleven years progress in earthworm ecotoxicology. Pedobiol 47:588–606.

Strong L. 1993. Overview: the impact of avermectins on pastureland ecology. Vet Parasitol 48:3–17.

Strong L, James S. 1992. Some effects of rearing the yellow dung-fly, *Scatophaga stercoraria* in cattle dung containing ivermectin. Entomol Exp Appl 63:39–45.

Svendsen C, Spurgeon DJ, Hankard PK, Weeks JM. 2004. A review of lysosomal membrane stability by neutral red retention: is it a workable earthworm biomarker? Ecotox Environ Saf 57:20–29.

Svendsen C, Weeks JM. 1997. Relevance and applicability of a simple biomarker of copper exposure. I. Links to ecological effects in a laboratory study with *Eisenia andrei*. Ecotox Environ Saf 36:72–79.

Tao S, Guo LQ, Wang XJ, Liu WX, Ju TZ, Dawson R, Cao J, Xu FL, Li BG. 2004. Use of sequential ASE extraction to evaluate the bioavailability of DDT and its metabolites to wheat roots in soils with various organic carbon contents. Sci of the Total Env 320:1–9.

ten Hulscher TEM, Postma J, den Besten PJ, Stroomberg GJ, Belfroid A, Wegner JW, Faber JM, van der Pol JJC, Hendriks AJ, van Noort PCM. 2003. Tenax extraction mimics benthic and terrestrial bioavailability of organic compounds. Env Tox and Chem 22:2258–2265.

Ter Laak TL, Mayer P, Busser FJM, Klamer HJC, Hermens JLM. 2005. A sediment dilution method to determine sorption coefficients of hydrophobic organic chemicals. Environ Sci and Technol 39:4220–4225.

Vale GA, Grant IF. 2002. Modelled impact of insecticide-contaminated dung on the abundance and distribution of dung fauna. Bull Entomol Res 91:251–263.

Van der Wal L, van Gestel CAM, Hermens JLM. 2004. Solid phase microextraction as a tool to predict interal concentrations of soil contaminants in terrestrial organisms after exposure to a standard laboratory soil. Chemosphere 54:561–568.

Volkering F, Quist JJ, van Velsen AFM, Thomassen PHG, Olijve M. 1998. A rapid method for predicting the residual concentration after biological treatment of oil polluted soil. In: Contaminated soil @98. Proc. of the Sixth International FZK/TNO Conference. London (UK): Thomas Telford, p 251–259.

Walker CH, Hopkin SP, Sibly RM, Peakall DB. 2001. Principles of ecotoxicology. New York (NY): Taylor & Francis.

Wardhaugh KG, Holter P, Longstaff BC. 2001. The development and survival of three species of coprophagous insect after feeding on the faeces of sheep treated with controlled-release formulations of ivermectin or albendazole. Aust Vet J 79:125–132.

Wardhaugh KG, Longstaff BC, Lacey MJ. 1998. Effects of residues of deltamethrin in cattle faeces on the development and survival of three species of dung-breeding insect. Aust Vet J 76:273–280.

Wardhaugh KG, Mahon RJ, Axelsen A, Rowland MW, Wanjura W. 1993. Effects of ivermectin residues in sheep dung on the development and survival of the bushfly, *Musca vetustissima Walker* and a scarabaeine dung beetle, *Euoniticellus fulvus Goeze.* Vet Parasitol 48:139–157.

Wiles JA, Krogh PH. 1998. Testing with the collembolans *I. viridis, F. candida* and *F. fimetaris.* In: Lokke H, van Gestel CAM, editors. Handbook of soil invertebrate toxicity tests. Chichester (UK): John Wiley, p 131–156.

Wratten SD, Forbes AB. 1996. Environmental assessment of veterinary avermectins in temperate pastoral ecosystems. Ann Appl Biol 128:329–348.

Yeates GW. 1994. Modification and qualification of the Nematode Maturity Index. Pedobiol 38:97–101.

8 Workshop Conclusions and Recommendations

Mark Crane, Katie Barrett, and Alistair Boxall

8.1 WORKSHOP CONCLUSIONS

The SETAC Workshop on Veterinary Medicines in the Environment concluded the following:

1) The impact of veterinary medicines on the environment will depend on several factors, including the amounts used, animal husbandry practices, treatment type and dose, metabolism within the animal, method and route of administration, environmental toxicity, physicochemical properties, soil type, weather, manure storage and handling practices, and degradation rates in manure and slurry.

2) The importance of individual routes into the environment for different types of veterinary medicines will vary according to the type of treatment and livestock category. Treatments used in aquaculture have a high potential to reach the aquatic environment. The main routes of entry to the terrestrial environment will be from the use of veterinary medicines in intensively reared livestock, via the application of slurry and manure to land, and the use of veterinary medicines in pasture-reared animals, where pharmaceutical residues will be excreted directly into the environment. Veterinary medicines applied to land by the spreading of slurry may also enter the aquatic environment indirectly via surface runoff or leaching to groundwater. It is likely that topical treatments will have a greater potential to be released to the environment than treatments administered orally or parenterally. Inputs from the manufacturing process, companion animal treatments, and disposal are likely to be minimal in comparison.

3) In contrast with substances that may be introduced directly into the environment, such as industrial chemicals, biocides, and pesticides, veterinary medicinal products are, in most cases, metabolized by animals (and may also be degraded in manure during storage time) before their introduction to the environment (exceptions are some aquaculture and ectoparasiticidal products). Thus, in addition to the medicine itself, its metabolites may enter and could affect the environment. Although most environmental impact assessments are based on the fate and effect

properties of only the parent substance, the environmental behavior of relevant metabolites should also be taken into consideration to predict if they would contribute to an increased overall risk to the environment. This may be achieved most cost-effectively by the use of quantitative structure-activity relationships (QSARs) and quantitative structure property relationships (QSPRs) if appropriate models can be developed for veterinary medicines. In addition to using QSAR and QSPR software tools, a significant amount of preliminary toxicity and safety information on many analogs of the medicinal product is already available during the discovery and predevelopment stages of a drug development program.

4) When a veterinary medicinal product contains more than one active ingredient, it might be relevant to base the risk assessment on not only the individual compounds but also their combination, especially when the compounds share the same mode of action. In such cases, the sum of the predicted environmental concentrations (PECs) of these active ingredients should be compared to the trigger value in VICH phase I in order to decide whether a phase II assessment is necessary.

5) Refinement of risk at higher tiers of risk assessment frameworks, such as those described in VICH guidance, usually involves a reduction in the conservatism of assumptions and an increase in realism, although single point estimates for the deterministic estimation of PECs and PNECs remain the norm. Sometimes increased realism might be achieved through the use of more realistic models of the environment, such as an estimation of a community NOEC from a mesocosm, or by the use of probabilistic risk assessment models.

6) To be effective, risk mitigation measures should meet the following criteria. They should a) reduce environmental exposure and transport of the veterinary medicine, b) be feasible with respect to agricultural practice, c) be consistent with applicable regulations, and d) have scientifically demonstrable effects.

7) Communication to the individuals responsible for carrying out the mitigation measure is often a significant challenge. An extensive communication strategy is needed to ensure that individuals are aware of their label responsibilities. Mitigation measures should be based on a realistic understanding of these communication challenges, including the background knowledge of the responsible individuals.

8) Useful feedback from pharmacovigilance may be weak because incident-reporting schemes can usually identify only gross examples of impacts.

9) The available methods for assessing aquatic exposures as a result of terrestrial applications of veterinary medicines generally provide conservative estimates of exposure concentrations, with some notable exceptions, such as strongly sorbed compounds.

10) Although a large body of data is now available on the transport of veterinary medicines into aquatic systems, much less information is available

on the fate and dissipation of veterinary medicines in receiving waters, with the exception of some aquaculture treatments.

11) The degradation processes in water or sediment may result in the formation of transformation products. The persistence and fate of these substances in surface water bodies may be very different from those of the parent compound. Current exposure assessment scenarios do not take into account the presence of metabolites or transformation products of veterinary medicines that could be biologically active.

12) Exposure assessments typically do not take into account ecosystem-level effects that occur as a result of multiple inputs of veterinary medicines. These scenarios are quite common, as intensive aquaculture and agricultural operations tend to be clustered in restricted geographical areas. Under these scenarios, inputs from multiple sources could be cumulative for exposures of aquatic organisms to waterborne contaminants. Exposure assessment methods are also not designed to assess mixtures of veterinary drugs. Assessments are typically conducted on the active ingredient(s) of a single veterinary medicinal product as part of an approvals process for its marketing. However, there is potential for mixtures of chemicals to impact aquatic organisms in an additive or greater than additive manner, especially when the veterinary medicines have similar mechanisms of action (e.g., antibiotics). These issues are particularly important when considering exposures to veterinary medicines that are marketed as mixtures, such as the potentiated sulfonamide antibiotics.

13) Veterinary medicines are biologically active substances, and there is an increasing body of evidence that a) exposure to select medicine groups may result in effects not identified using standard methodologies and b) indirect effects may be elicited. Moreover, the exposure profiles and bioavailability of veterinary medicines in the natural environment will likely be very different than in the laboratory. By combining information on a substance's pharmacology and toxicology in target organisms and humans with ecotoxicogenomic approaches and higher tier assessment approaches developed for pesticides, it should be possible to develop a much better understanding of the real risks of veterinary medicines to aquatic systems. Many of these approaches also have potential applications in retrospective assessment work such as postauthorization monitoring, watershed assessments, and toxicant identification evaluations.

14) No validated or standardized method for assessing the fate of veterinary medicines in manure at either the laboratory level or field level exists, and tests in existing pesticide or OECD guidelines do not cover these aspects. In terms of fate we have poor knowledge of what happens in slurry prior to soil amendment, but this is an important area for risk management.

15) In many confined animal and poultry production systems, waste is stored for some time, during which time a transformation of veterinary medicines could occur prior to release of material into the broader environment. Manure-handling practices that could accelerate veterinary medicine

dissipation — for example, composting — offer an opportunity to reduce environmental exposure significantly. There are no standardized methods for evaluating the fate of pharmaceuticals in manure and very little published information on fate characteristics during manure storage.

16) An assessment of a medicine's potential to affect the terrestrial and aquatic environment negatively is not evaluated in isolation. The data package used to assess the efficacy and safety of a veterinary medicine under development is extensive. Safety data packages for medicines intended for livestock include the results of studies to test the safety of the medicine in the target animal species. Target livestock species are typically cattle, pigs, and poultry. Toxicity data are used to evaluate the safety to the consumer of ingestion of animal tissues (e.g., muscle, kidney, liver, or milk) containing medicine residues (human food safety). Furthermore, an evaluation is conducted to determine the potential impact of veterinary medicine residues on the normal gastrointestinal tract flora of humans (microbial safety). Finally, data from toxicity studies are used to address whether the farmer should be concerned for his or her safety when the medicine is administered to the target animal species (user safety). All of these data should be leveraged for use in the ecotoxicity assessment.

8.2   WORKSHOP RECOMMENDATIONS

The following recommendations were made by the workshop:

1) Usage data are unavailable for many groups of veterinary medicines and for several geographical regions, which makes it difficult to establish whether these substances pose a risk to the environment. It is therefore recommended that usage information be obtained for these groups, including the antiseptics, steroids, diuretics, cardiovascular and respiratory treatments, locomotor treatments, and immunological products. Better usage data will assist in designing more robust hazard and risk management strategies that are tailored to geographically explicit usage patterns.

2) From the information available, it appears that inputs from aquaculture and herd or flock treatments are probably the most significant in terms of environmental exposure. This is mainly because many aquaculture treatments are dosed directly into the aquatic environment, and herd or flock treatments may be excreted directly onto pasture. However, the relative significance of novel routes of entry to the environment from livestock treatments, such as washoff following topical treatment, farm yard runoff, and aerial emissions, has not generally been considered. For example, the significance of exposure to the environment from the disposal of used containers or from discharge from manufacturing sites should be investigated further. In addition, substances may be released to the environment

as a result of off-label use and poor slurry management practice. The significance of these exposure routes is currently unknown.

3) Environmental risk assessment is unlike human or target species risk assessment because of the much wider range of species and exposure pathways that must be considered. This makes accurate prospective risk assessment difficult at the authorization stage. Therefore, a regulatory scheme that does not involve credible postauthorization monitoring is likely to suffer from an unknown number of false negatives, in which the environmental risks of chemicals are underestimated. There is a need, therefore, for more active strategic monitoring of the environmental fate and effects of those veterinary medicines that have the potential to cause harm to the environment.

4) The predicted concentrations of strongly sorbing antibiotics such as tetracyclines in surface water and groundwater tend to be underestimated, as the models do not consider colloidal or particle-bound transport. Studies to investigate the mechanisms of transport of highly sorbing substances and subsequent model refinements are therefore warranted.

5) In the case of aquatic exposure assessments related to aquaculture facilities, the available assessment methods require further development. More sophisticated exposure models are required, especially in the case of intensive net pen aquaculture. Exposure scenarios for different aquaculture systems (pond, net pen, flow-through, etc.) for specific applications of medicines (bath versus feed) are needed. Operational data are also needed for the aquaculture facilities to refine the exposure scenarios (e.g., flow rates used, dilution factors, and number of treatments). Additional monitoring data are needed to examine the appropriateness of the aquaculture exposure scenarios for screening-level risk assessments.

6) Research is required to improve our understanding of the relative importance of partitioning processes for drugs (in water, feces, etc.), degradation processes, and other dissipation mechanisms in order to determine the most appropriate way to calculate PECs for aquatic systems. As inputs are likely to be intermittent or pulsed for some medicines (e.g., bath treatments), more consideration should also be given to approaches that link the temporal variability of aquatic exposures to effects, such as the use of time-weighted averages.

7) In terms of risk management, more work needs to be done to identify beneficial management practices (BMP) that can be used to mitigate exposures of aquatic organisms. So far there have been hardly any studies to evaluate the capacity of BMPs such as the use of optimized tillage practices and the maintenance of buffer strips and riparian zones to reduce aquatic exposures from the terrestrial application of veterinary medicines. In the case of current use pesticides, there is ample evidence that inputs into aquatic systems can be mitigated by the use of these BMPs.

8) Our current understanding of certain areas of aquatic effects assessment is poorly developed, and future efforts should focus on a number of key

areas, namely, a) the accumulation of data and knowledge to test and further refine extrapolations from mammalian toxicity data to aquatic effects, b) the further development of ecotoxicogenomic approaches and exploration of how data from these can be applied in risk assessment, c) the development and validation of methods for metabolite and degradate assessment, d) studies to understand further those factors and processes affecting the bioavailability and trophic transfer of veterinary medicines in aquatic systems, and e) consideration of the potential impacts of mixtures of veterinary medicines and mixtures containing veterinary medicines and other contaminant classes.

9) There should be development of clear guidance specific to veterinary medicines for laboratory and field-based methods for the evaluation of degradation and dissipation. These should take into account agronomic practice when appropriate (e.g., the addition of manure or slurry). The impact of different storage and composting conditions on the degradation of veterinary medicines needs to be better understood and investigated. There is very little knowledge of the dissipation kinetics and transformation pathways for veterinary medicines in manures stored under commercial conditions. This information is required to improve estimates of PEC_{soil} and to validate manure storage BMPs (e.g., composting) with respect to reducing veterinary medicine concentrations. We recommend that systematic experimental determination of veterinary medicine persistence in appropriate manures incubated under realistic conditions should be performed.

10) Field-based validation of PEC modeling methods needs to be conducted, as there is a perception that existing methods may be too conservative and unrealistic.

11) Exposure scenarios following the application of combination products need to be considered.

12) The development of tier A dung fauna toxicity-testing methods has been in progress for some years under the auspices of the SETAC Dung Organism Toxicity Testing Standardization (DOTTS) group. Although the development of these methods has been given a high priority by the OECD, only a limited number of laboratories are participating in the ring testing and only limited man hours allocated to the testing effort have been possible as the work has no funding. This initiative should be supported more fully. Alternatively, a simple model may be a valuable tool for use in risk assessment and management for dung fauna.

13) Modeling of population and ecosystem effects, alternative endpoints (e.g., biomarkers), and the biological relevance of bound residues should all be investigated further.

Index

A

ACRs, *see* Acute:chronic ratios
Active pharmaceutical ingredient (API), 11
Acute:chronic ratios (ACRs), 101
ADME, *see* Adsorption, distribution,
    metabolism, and elimination
Adsorption, distribution, metabolism, and
    elimination (ADME), 1
  bioaccumulation potential, 105
  degree of metabolism and, 130
  environmental risk assessment and, 22
  profiles, target animal, 116
  studies
    aquatic effects assessment, 102
    rate of metabolism for confined animals
      in, 34
  toxicological endpoints, 104
AF, *see* Assessment factor
AHI, *see* Animal Health Institute
Albendazole, 131
Aleochara spp., 160
Allium cepa, 143
Altrenogest, 9
Aminoglycosides, 8, 9, 130, 135
Amoxicillin, 9
Ampicillin, 9
Amprolium, 9, 10
Anaesthetics, 9
Analgesics, 9
Animal Health Institute (AHI), 145
Antibacterials, 8, 9
Antibiotics
  aquaculture applications, 62
  European regulation of, 23
  food digestion and, 10
  major route of entry into environment, 13,
    129
  number of tons used by European Union, 8
  persistence in manure, 135
  plant absorption of, 143
  potentiated sulfonamide, 183
  predicted concentrations, 185
  soil-adsorbed antibiotics, 142
  tetracycline, interaction with organic
    matter, 64
  total usage in United Kingdom, 22
  uses, 21

Antifungals, 9, 10
Antiprotozoals, 10
Aphodius spp., 38
API, *see* Active pharmaceutical ingredient
Apramycin, 9
Aquaculture
  antibiotics used in, 62
  BMPs in, 90–91
  freshwater, 76
  labeling for, 43
  major route for environmental
    contamination, 12
  medicines, 10
  PEC values, 77
  ponds, operation as open systems, 78
  treatments, 9
  VMPs, 29
Aquatic hazards of veterinary medicines,
    assessment of, 97–128
  acute-to-chronic extrapolation, 110
  antimicrobials, 104
  application factors and species sensitivities,
    110–113
  binning of chemicals with similar MOAs,
    117
  CAFO-impacted watershed, 117
  current methods of aquatic effects
    assessment, 98–102
    higher tier testing, 99–100
    limitations, 101
    lower tier approaches, 99
  drug–drug interaction profiles in mammals,
    117
  effects of veterinary medicines in natural
    environment, 113–121
    assessing effects on communities,
      119–121
    enantiomer-specific hazard, 117–118
    episodic exposures, 114
    matrix effects, 114–115
    metabolites and degradates, 115–116
    mixtures, 116–117
    sorption to sediment, 118–119
  example scenarios, 101
  feed utilization efficiency, promotion of,
    104
  fingerprinting, 109

generation of toxicity data, 110
genomic techniques, 108
global information for organism, 108
hormone-mediated effects, 104
mammalian toxicology studies, 104
multispecies responses, 119
narcosis MOA, 115
novel approaches to aquatic effects
    assessment, 102–110
    chemical characteristics, target
        organism efficacy data,
        toxicokinetic data, and mammalian
        toxicology data, 102–107
    ecotoxicogenomics, 108–110
physicochemical characteristics of
    emamectin benzoate, 106
PNECs for aquatic organisms exposed to
    antibiotic, 111
profiling, 109
protection goals, 98
results from mammalian tests used to target
    environmental effects testing, 103
safety factors, 110
screening assessment approach to target
    aquatic effects testing with fish
    from water exposure, 105
species sensitivity distributions for aquatic
    organisms exposed to antibiotic in
    water, 112
tier B tests, 100
types and characteristics of cosms, 120
unknown variability, 110
Aquatic systems, exposure assessment of
    veterinary medicines in, 57–95
aquaculture treatments (experimental
    studies), 73–89
    freshwater aquaculture, 76
    inputs and fate of marine aquaculture
        treatments, 75–76
    modeling exposure from aquaculture
        treatments, 77–89
aquatic exposure to veterinary medicines
    used to treat livestock (experimental
    studies), 62–73
    comparison of modeled concentrations
        with measured concentrations,
        66–73
    leaching to groundwater, 63
    movement to surface water, 63–64
    predicting exposure, 65–66
calculation shell, 66
chemical patch, 87
control of sea lice infestations, 62
CVMP guidance document, 66
default assumption, 85
default dilution factor, 78

deposition footprint, 88
environmental introduction concentrations,
    77
excreted oxytetracycline partitions, 76
exposure scenario evaluation, 57
fish kills, 59
marine net pens, PEC values, 77
medicated feeds, 78, 79
model assessment for tylosin, 70
model of atrazine transport, 90
open ponds, calculations for, 78
open water systems, 81
recirculating systems, 86
release of chemotherapeutic agents, 62
sea lice treatments, 75
sheep dip products, 59
sources of veterinary medicines, 58–62
    treatments used in agriculture, 58–61
    treatments used in aquaculture, 61–62
SWASH, 66
tetracycline antibiotics, interaction with
    organic matter, 64
time-averaged PECs, 84
US government hatchery facilities, 85
whole-pond bath treatments, 78
worst-case estimate, 88
Arable lands, 157
Assessment factor (AF), 31, 110
applicability of in VICH approach, 112
plant, 31
safety factors, 110
VICH, 171
Atrazine transport, model of, 90
Avermectin compounds, 75
Azamethiphos, 75
Azole, 9, 130

B

Bacillus subtilis, 111, 112
BAFs, *see* Bioaccumulation factors
Benzimidazoles, 9, 134
Benzylpenicillin, 9
Best management practices (BMP), 90, 136
    aquaculture, 90–91
    aquatic organisms, 90, 185
    manure type characteristics and, 136
Beta-lactams, 8, 9, 135
Biguanide/gluconate, 9
Bioaccumulation factors (BAFs), 107
Bioindicator(s)
    indicator species as, 162
    responses, as risk assessment tools, 175
    systems, monitoring of manure-treated
        areas, 176

BMPs, *see* Best management practices
Brassica oleracea, 143

C

CAFO, *see* Concentrated animal-feeding
    operation
Caprofen, 9
Ceftiofur, 64
Center for Veterinary Medicine (CVM), 145
Cephalosporins, 8
Ceriodaphnia dubia, 118
Chemotherapeutants, 12
Chitin synthesis inhibitors, 75
Chlorhexidine, 9, 10
Chlortetracycline, 9, 143
Clopidol, 9, 10
CMC, *see* Criterion maximum concentration
Coccidiostats, 9, 10
Committee for Medicinal Products for
    Veterinary Use (CVMP), 65, 146
  algorithms suggested by, 67–70
  guidance document, 66
  guideline, 146
  model, 148
Companion animal(s), 3, 14
  products, 15
  source of veterinary medicines from, 59,
    60–61
  treatments, 16, 21, 57, 181
Concentrated animal-feeding operation
    (CAFO), 50
  -impacted watershed, 117
  lagoons, runoff from, 50
  pulsed exposures, 114
  transport of veterinary medicines from, 114
Cosms, types and characteristics, 120
Criterion maximum concentration (CMC), 114
CVM, *see* Center for Veterinary Medicine
CVMP, *see* Committee for Medicinal Products
    for Veterinary Use
Cyclodextrin extraction, 175
Cypermethrin, 9, 75, 131

D

Daphnia magna, 31, 100, 111, 118
Decoquinate, 10
Deltamethrin, 9, 75, 131
2,4-Diaminopyrimidines, 9
Diazinon, 9, 131, 134
Diclazuril, 9, 10
Diflubenzuron, 75
Digestive enhancers, 10

Dihydrostreptomycin, 9
Dimethicone, 9
Dissolved organic carbon (DOC), 86
DOC, *see* Dissolved organic carbon
Doramectin, 8, 9, 131, 134
DOTTS group, *see* Dung Organism Toxicity
    Testing Standardization group
Drug development program, preliminary
    toxicity information available
    during, 35
Dung Organism Toxicity Testing Standardiza-
    tion (DOTTS) group, 167–168, 186

E

EA, *see* Environmental assessment
Earthworm toxicity tests, artificial soil in, 36
Ecotoxicological soil quality criteria (ESQC),
    111
Ectoparasiticide(s), 8, 21
  applied externally to canine species, 15
  arsenic-based, 46
  disposal, 15
  dosing regimes, 45
  pasture invertebrates and, 28
  pour-on formulations of, 32
  reference organophosphate, 109
  sheep dip, 15
  target species, 37
  theoretical, 38
  types, 8
  uses, 8
EIS, *see* Environmental impact statement
Emamectin benzoate, 9, 75, 106
Enrofloxacin, 9
Enteric bloat preps, 9
Environmental assessment (EA), 24
Environmental impact statement (EIS), 24
Environmental Quality Standard (EQS), 14
Environmental risk assessment and manage-
    ment of veterinary medicines, 21–55
  arsenic-based ectoparasiticides, 46
  criteria for classifying effects of veterinary
    medicines in ecosystem, 42
  distribution of effect values in simple
    probabilistic model of dung insect
    toxicity, 40
  extensively metabolized medicine, 27
  green labeling of products, 45
  labels for persistent products, 43
  metabolite evaluation, 34
  MRL violations, 47, 49
  natural systems, 50
  North Bosque River watershed, 51
  oxpecker-compatible dips, 46

parameters for estimating parasiticide
impacts on dung insect populations,
40
pass–fail threshold outputs, 41
preliminary toxicity information, 35
refinement of veterinary medicinal product
risk assessments, 33–41
combination products, 35–36
metabolism and degradation, 33–35
probabilistic risk assessment of
veterinary medicines, 36–41
refinement of environmental exposure
predictions, 36
regulatory perspective, 23–33
Australia, 26
Canada, 26–27
current guidelines, 27–33
European Union, 25–26
Japan, 26
legislation, scope, and past guidelines
for environmental risk assessment,
23–27
overview, 24
United States, 24–25
VICH and VICH–EU technical
guidance document, 27–33
restricted spreading of manure, 43
risk management, 41–52
communication challenge, 44–47
incidence reporting and
pharmacovigilance, 47–49
postmarket monitoring and
remediation, 51–52
retrospective risk assessment, 49–51
risk assessment and management
beyond authorization or approval,
44–49
risk mitigation measures within product
authorization or approval, 42–44
surveillance scheme, 51
temporal distribution of main seasonal
activity of *Aphodius* spp., 39
VICH phase decision tree, 28, 29
VICH phase II guidance, 29, 30
VICH tier B effects studies, 31
Environmental risk limits (ERLs), 111
Eprinomectin, 8, 9, 131
EqP, *see* Equilibrium partitioning theory
EQS, *see* Environmental Quality Standard
Equilibrium partitioning theory (EqP), 119
ERLs, *see* Environmental risk limits
Erythromycin, 9
Escherichia coli, 142
ESQC, *see* Ecotoxicological soil quality
criteria

Estradiol, 109
Estradiol benzoate, 9
Euoniticellus intermedius, 163
Euthanasia products, 9

F

FA, *see* Fluctuating asymmetry
Farm environment, *see* Terrestrial systems,
exposure assessment of veterinary
medicines in
FDA, *see* US Food and Drug Administration
Feed(s)
additives, dietary-enhancing, 7
medicated, 78, 79
chemotherapeutics in, 62
control of sea lice by, 75
drugs administered as, 12
fate processes, 88
flow-through scenario, 85
uneaten, 82
utilization efficiency, promotion of, 104
waste, 12, 156
Feedstuffs
antibiotic compounds added to, 10
antiprotozoals incorporated into, 10
Fenbendazole, 9, 131
Finding of no significant impact (FONSI), 25
Fish farm(s)
characteristics, deposition footprint and, 88
chemotherapeutic medicines used in, 12
use of medicines in, 7
Flavophospholipol, 9
Florfenicol, 9
Fluctuating asymmetry (FA), 175, 176
Fluoroquinolones, 8, 9, 130
FOCUS, *see* Forum for the Coordination of
Pesticide Fate Models and Their Use
FOCUS model, 65, 66
Folsomia
candida, 166
fimetaria, 166
FONSI, *see* Finding of no significant impact
Forum for the Coordination of Pesticide Fate
Models and Their Use (FOCUS), 66
Furazolidone photodegradation, 64

G

Good laboratory practice (GLP), 35
Green labeling, 45
Griseofulvin, 9, 10
Growth promoters, 9, 10–11

H

Halothane, 9
Hazard quotient (HQ), 99
HC5, 171
High production volume (HPV) chemicals, 22
Hormone(s), 9, 10
  functions regulated by, 21
  major route of entry into environment, 13,
    129
  model, 109
  modulation, mode of action via, 100
  restricted uses of, 10
HPV chemicals, see High production volume
    chemicals
HQ, see Hazard quotient
Human medicines, exposure of wildlife to, 1
Hydrogen peroxide, 9, 75

I

Indicator species
  as bioindicators in field situation, 162
  information provided by, 162
  standard guidelines, 160
Inputs of veterinary medicines, see Uses and
    inputs of veterinary medicines in
    environment
International Standards Organization (ISO), 29
International Union of Pure and Applied
    Chemistry (IUPAC), 142
Ionophores, 8
ISO, see International Standards Organization
Isoflurane, 9
IUPAC, see International Union of Pure and
    Applied Chemistry
Ivermectin, 8, 9, 64, 131, 132, 134

L

Lasalocid, 9
Lasalocid acid, 10
LC-MS-MS analysis, see Liquid
    chromatography tandem mass
    spectrometry analysis
Lepomis macrochirus, 111
Levamisole, 9, 131
Lidocaine, 9
Lignocaine, 9
Lincomycin, 9, 64
Lincosamides, 130
Liquid chromatography tandem mass spec-
    trometry (LC-MS-MS) analysis, 34
Litopenaeus vannamei, 111

Livestock
  characteristics, 65
  control of external parasites in, 8
  distribution routes for veterinary
    medicines, 59
  drug–drug interaction profiles, 117
  ectoparasites of, 37
  facilities, intensive, 114
  grazing, 158
  manure, arthropods associated with, 160
  medicines, aquatic exposure to, 62
  production, 13–14
    pastures for, 158
    protection of native biodiversity versus,
      167
  safety data packages for medicines
    intended for, 156
  toxicity data, 184
Lysosomal membrane stability, 175

M

Macrocyclic lactones, 130
Macrolide endectins, 9, 130
Macrolides, 9, 63, 130, 135
Maduramicin, 9, 10
MAH, see Marketing authorization holder
Management, see Environmental risk
    assessment and management of
    veterinary medicines
Manure
  application, release of veterinary medicines
    via, 7
  arthropods associated with, 160
  characteristics, 136
  degradation products formed in, 172
  disposal of onto land, 13, 129
  distribution, PEC refinement, 148
  earthworms found in, 165
  -handling practices, veterinary medicine
    dissipation and, 134–135, 183–184
  management practices, model, 65
  microbiological degradation pathways and,
    140
  monitoring of medicine effects in areas
    treated with, 176
  nitrogen or phosphorus content of, 144
  plant absorption of antibiotics present in,
    143
  predicted default concentrations, 67
  spreading, regulations, 43
  storage, 173
    fate during, 134
    scenario, 70

transport, fish kills and, 59
withdrawal period, 44
Marketing authorization holder (MAH), 47
Maximum inhibition concentrations (MIC), 112
Maximum residue level (MRL)
    regulations, 44
    violations, 47, 49
Measured environmental concentrations (MECs), 66
MECs, *see* Measured environmental concentrations
Medroxyprogesterone, 9
Mesocosm(s)
    characteristics, 120
    radiolabeled chemicals used in, 170
    stream, experiments, 121
    study, community NOEC from, 36, 182
    terrestrial, multiple species effects, 169
Metamyzole, 9
Methyltestosterone, 9
Metronidazole, 140
MIC, *see* Maximum inhibition concentrations
Miconazole, 9, 10
MOA, *see* Mode of action
Mode of action (MOA), 108
    binning of chemicals with similar, 117
    fingerprint data library, 109
    narcosis, 115
    -oriented toxicology studies, 109
    toxic, identification of, 108
    uncertainties, 108
Model(s)
    aquatic systems
        atrazine transport, 90
        exposure concentrations in soil, 65
        longer-term, 87
        short-term, 87
    assessment for tylosin, 70
    concentration of veterinary medicine in soil, 143
    CVMP, 148
    drug products, 35
    environment, realistic, 36
    environmental risk assessment, 35
    EqP, 119
    exposure, 32
    FOCUS, 65, 66
    hormones, 109
    impacts of veterinary medicines to fish, 97
    manure management practices, 65
    metabolites, 115
    nonmammalian, 109
    parasiticides, 38
    PEC calculation, 32
    pesticide root zone, 66

probabilistic, dung insect toxicity, 40
QSAR, toxicity pathway, 115
sensitivity, 40
stream systems, 121
VetCalc, 65, 70
watershed, 51
Monensin, 9, 64
Morantel, 8, 9, 131
Moxidectin, 8, 131
MRL, *see* Maximum residue level

N

Narasin, 9, 10
National Environmental Policy Act of 1969 (NEPA), 24
Natural protected systems, 158
Natural systems, description of, 50
Navicula pelliculosa, 111
Nematode Maturity Index (NMI), 162
Neomycin, 9
NEPA, *see* National Environmental Policy Act of 1969
Nicarbazin, 9, 10
NMI, *see* Nematode Maturity Index
NOECs, *see* No observed effect concentrations
Nonsteroidal anti-inflammatory drugs (NSAIDs), 9, 11
No observed effect concentrations (NOECs), 99
NSAIDs, *see* Nonsteroidal anti-inflammatory drugs

O

Olaquindox, 140
Oncorhynchus mykiss, 111
Open systems
    aquaculture, 77, 78
    VICH phase II, 27
Organophosphates, 9, 75
Oxfendazole, 8, 131
Oxytetracycline, 9, 62, 64, 76

P

PAF, *see* Potentially affected fraction
Parasiticides, 8, 9, 38
PEC, *see* Predicted environmental concentration
PEHA, *see* Probabilistic ecological hazard assessment
Penicillin G, 9
Pentobarbitone, 9
Permethrin, 8

Pesticide(s)
Australian assessment of environmental
    risk, 26
bound residues, 141–142
current-use, BMP, 90
degradates, 115
dissipation, 137
leaching behavior of, 66
metabolites, 142
physicochemical techniques tested with, 175
registration, 108, 121
regulation, 111
risk, laypeople ranking of, 45
root zone model (PRZM), 66
scenarios, 140
secondary poisoning and, 173
total amount used in United Kingdom, 22
transport from agricultural fields, 62
use of farm ditches in removing, 90
Pets, *see* Companion animal
Phenobarbitone, 9
Phenylbutazone, 9
Pimephales promelas, 118
Pleuromutilins, 9
PNECs, *see* Predicted no-effect concentrations
Pollutant
    incidents, notification of, 47
    nutrient, nonpoint source, 50
    retention, 175
    sheep treatment, 13, 14
Poloxalene, 9
Potentially affected fraction (PAF), 172
PRA, *see* Probabilistic risk assessment
PRCs, *see* Principal response curves
Predicted environmental concentration (PEC),
    28, 163–164, 182
Predicted no-effect concentrations (PNECs),
    99, 162
Principal response curves (PRCs), 171
Probabilistic ecological hazard assessment
    (PEHA), 113
Probabilistic risk assessment (PRA), 36, 37
Procaine, 9
Progesterone, 9
PRZM, *see* Pesticide root zone model
Pseudokirchneriella subcapitata, 111
Pyrantel, 9
Pyrethroids, 9, 75
Pyrimidines, 9

Q

QSAR, *see* Quantitative structure-activity
    relationship
Quantitative structure-activity relationship
    (QSAR), 2, 182

bioaccumulation potential, 2
models, toxicity pathways, 115
prediction of chemical properties using, 35
Quinolone photodegradation, 64

R

Rapid persulfate oxidation, 175
Risk assessment, *see* Environmental risk
    assessment and management of
    veterinary medicines
Robenidine, 9
Robenidine hydrochloride, 10

S

Salinomycin, 9
SARSS, *see* Suspected Adverse Reaction
    Surveillance Scheme
Scatophaga stercoraria, 175
Secondary poisoning, 173
Semipermeable membrane devices, 175
SETAC Workshop on Veterinary Medicines in
    the Environment, 2
    conclusions, 181–184
    objectives, 2
    recommendations, 184–186
Sheep-dipping activities, 13
Sites of special scientific interest (SSSI), 158
Skeletonema costatum, 110, 111
Slurry
    characteristics, 136
    contamination from, 158
    disposal of onto land, 13
    intensively reared animals and, 144
    pig, 64, 76
    poor management, 184–185
    sorption behavior of veterinary medicines,
        141
    storage, optimum period for, 169
    stored, aerated, 135
    timing of application, 138
    transport of sulfonamides in, 63
    waste stored as, 134
Soil
    -adsorbed antibiotics, 142
    artificial, earthworm toxicity tests, 36
    concentration of veterinary medicine in,
        model, 143
    exposure concentrations in, aquatic
        systems and, 65
    organic matter, propensity for sorption to,
        129
    water and assessment tool (SWAT), 51

Solid-phase micro extraction, 175
Soluble reactive phosphorus (SRP), 50–51
Solvent extraction techniques, 175
Species sensitivity distribution (SSD), 110
　aquatic organisms exposed to an antibiotic
　　in water, 112
　PNEC derived from, 171
　safety factors and, 170
Spiramycin, 9
SRP, *see* Soluble reactive phosphorus
SSD, *see* Species sensitivity distribution
SSSI, *see* Sites of special scientific interest
Stocking density, 148
Stream systems, model, 121
Sulfadiazine, 9, 64
Sulfamethazine, 9, 64
Sulfathiazole, 9
Sulfonamides, 8, 63, 135
Sulphonamides, 9, 130
Supercritical fluid extraction technique, 175
Surface water-sampling programs, European, 1
Surface water scenarios help (SWASH)
Surfactant extraction, 175
Suspected Adverse Reaction Surveillance
　　Scheme (SARSS), 47
SWASH, *see* Surface water scenarios help
SWAT, *see* Soil water and assessment tool

T

Teflubenzuron, 9, 75
Tenax extraction, 175
Terrestrial environment, assessing of effects of
　　veterinary medicines on, 155–180
　alternative endpoints, 175–176
　bioavailability to soil organisms, 174
　bioindicator approaches, 176–177
　bioindicator responses, 175
　bound residues, 174–175
　calculation of PNEC concentrations and
　　use of assessment factors, 171–172
　categories of land use, 157
　cellular marker for stress, 175
　considerations unique to veterinary
　　medicines, 155–156
　　additional safety data available in
　　　dossier, 156
　　residue data and detoxification by target
　　　animal species, 156
　　routes of entry, 155–156
　coprophilous insects, 166
　cyclodextrin extraction, 175
　environmental profile, 158
　functional groups, 167
　fungivorous beetles, 166

HC5, 171
hockey stick-shaped degradation curve, 174
justification for existing testing methods, 160
metabolite testing in tiers A and B, 172–173
migratory species, 157
modeling population and ecosystem effects,
　176–177
Nematode Maturity Index, 162
on-site species, monitoring of, 170
organism behavior, 174
pasture management, 167
pollutant retention, 175
protection goals, 157–159
rainfall variation, 170
rapid persulfate oxidation, 175
research needs, 177
secondary poisoning, 173
semipermeable membrane devices, 175
sequestration of hydrophobic pollutants, 174
short-term and sublethal effects tests, 163
solid-phase micro extraction, 175
solvent extraction techniques, 175
species indicator applications, types of, 162
supercritical fluid extraction technique, 175
surfactant extraction, 175
Tenax extraction, 175
tier A testing, 163–168
　collembolans, 166
　dung fauna, 166–168
　earthworms, 165
　fate, 164
　microorganisms, 164–165
　physicochemical properties, 163–164
　plants, 165
tier B testing, 168–169
tier C testing, 169–171
　mesocosm and field testing, 169–170
　monitoring studies, 170–171
　testing of additional species, 170
toxicity data, 156
use of indicator species, 160–162
Terrestrial systems, exposure assessment of
　　veterinary medicines in, 129–153
absorption and excretion by animals,
　130–134
acid–base reflux procedures, 142
active ingredients used in veterinary
　products 139
commonly employed practices for manure
　storage and handling, 135
comparison of predicted environmental
　concentration in soil, 147
concentrations of chlortetracycline in plant
　tissues, 143
degree of metabolism of major therapeutic
　classes of veterinary medicines, 130

emission scenarios, 143
excretion profiles of ivermectin, 132
factors affecting dissipation in farm
    environment, 137–142
    abiotic degradation processes, 140–141
    biotic degradation processes, 138–140
    bound residues, 141–142
    dissipation and transport in dung
        systems, 137–138
    dissipation and transport in soil
        systems, 138–141
    sorption to soil, 141
fate during manure storage, 134–136
laboratory degradation studies of active
    substances in soil, 135
manure type characteristics, 136
measured and predicted environmental
    concentrations for range of
    veterinary medicines, 147
metabolism data, 130
models for estimating concentration of
    veterinary medicine in soil, 143–148
    intensively reared animals, 144–147
    pasture animals, 148
    PEC refinement, 148
parasiticide formulations available in
    United Kingdom, 131
percentage of applied dose excreted in dung
    and urine as parent molecule and/or
    metabolites, 133
releases to environment, 136–137
research needs, 149
stocking density, 148
terrestrial exposure assessments, 143
time-dependent sorption, 142
total residue approach, 148
uptake by plants, 143
Tetracyclines, 8, 9, 63, 64, 130, 142
Tiamulin, 9
TMDL, *see* Total maximum daily load
Toltrazuril, 9, 10
Total maximum daily load (TMDL), 50, 51
Total residue approach, 148
Tranquilizers, 9
Trenbolone, 109
Triclabendazole, 9, 131
Trimethoprim, 9, 63, 64
Tylosin, 9, 70, 140

U

Uses and inputs of veterinary medicines in
    environment, 7–20
    aquaculture medicines, 10
    available data, 7

digestive enhancers, 10
landfilling, 15
major groups of veterinary medicines, 9
major usage protozoal compounds, 10
pathways to environment, 11–15
    agriculture (livestock production),
        13–14
    aquaculture, 12
    companion and domestic animals,
        14–15
    disposal of unwanted drugs, 15
    emissions during manufacturing and
        formulation, 11–12
    regulatory requirements, 7
    sheep-dipping activities, 13
    summary, 16–17
    topical applications, environmental
        contamination by, 13, 16
    veterinary medicine use, 7–11
        antibacterials, 8
        antifungals, 10
        coccidiostats and antiprotozoals, 10
        growth promoters, 10–11
        hormones, 10
        other medicinal classes, 11
        parasiticides, 8, 9
US Food and Drug Administration (FDA), 23,
    24, 25, 100

V

VetCalc model, 65, 70
Veterinary medicinal product (VMP), 28, 33
    aquaculture, 29
    compilation of pharmacovigilance data, 49
    marketing authorization, 49
    refinement of risk assessments, 33–41
        combination products, 35–36
        environmental exposure predictions, 36
        metabolism and degradation, 33–35
        probabilistic risk assessment, 36–41
    reporting of adverse event, 48
Veterinary Medicines Directorate (VMD), 7
VICH
    applicability of AFs in, 112
    assessment factors, 171
    Ecotoxicity Guideline, 27
    guidance
        scenario, 60
        tiered testing strategy, 160
    guidelines
        ecotoxicity studies, 172
        tier C studies, 169
    phase I, 99
    phase II, 155, 164

Steering Committee, 27
tier B tests, 100
trigger value, 88
VMD, *see* Veterinary Medicines Directorate
VMP, *see* Veterinary medicinal product

W

Watershed
assessments, retrospective, 183
CAFO-impacted, 117
characteristics, PEC value adjustment and,
80–81
fate and transport models, 51

model, 51
North Bosque River, 50, 51
pond
with bath treatment, 78
medicated feeds in, 79
treatment interval and, 86
soluble reactive phosphorus in, 50–51
Workshop, *see* SETAC Workshop on
Veterinary Medicines in the
Environment

Z

Zea mays, 143

Other Titles from the Society of Environmental Toxicology and Chemistry (SETAC)

Freshwater Bivalve Ecotoxicology
Farris and Van Hassel, editors
2006

Estrogens and Xenoestrogens in the Aquatic Environment:
An Integrated Approach for Field Monitoring and Effect Assessment
Vethaak, Schrap, de Voogt, editors
2006

Assessing the Hazard of Metals and Inorganic Metal Substances in Aquatic and Terrestrial Systems
Adams and Chapman, editors
2006

Perchlorate Ecotoxicology
Kendall and Smith, editors
2006

Natural Attenuation of Trace Element Availability in Soils
Hamon, McLaughlin, Stevens, editors
2006

Mercury Cycling in a Wetland-Dominated Ecosystem: A Multidisciplinary Study
O'Driscoll, Rencz, Lean
2005

Atrazine in North American Surface Waters: A Probabilistic Aquatic Ecological Risk Assessment
Giddings, editor
2005

Effects of Pesticides in the Field
Liess, Brown, Dohmen, Duquesne, Hart, Heimbach, Kreuger, Lagadic, Maund, Reinert, Streloke, Tarazona
2005

Human Pharmaceuticals: Assessing the Impacts on Aquatic Ecosystems
Williams, editor
2005

Toxicity of Dietborne Metals to Aquatic Organisms
Meyer, Adams, Brix, Luoma, Stubblefield, Wood, editors
2005

Toxicity Reduction and Toxicity Identification Evaluations for Effluents, Ambient Waters,
and Other Aqueous Media
Norberg-King, Ausley, Burton, Goodfellow, Miller, Waller, editors
2005

Use of Sediment Quality Guidelines and Related Tools for the Assessment of Contaminated Sediments
Wenning, Batley, Ingersoll, Moore, editors
2005

Life-Cycle Assessment of Metals
Dubreuil, editor
2005

Working Environment in Life-Cycle Assessment
Poulsen and Jensen, editors
2005

SETAC

A Professional Society for Environmental Scientists and Engineers and Related Disciplines Concerned with Environmental Quality

The Society of Environmental Toxicology and Chemistry (SETAC), with offices currently in North America and Europe, is a nonprofit, professional society established to provide a forum for individuals and institutions engaged in the study of environmental problems, management and regulation of natural resources, education, research and development, and manufacturing and distribution.

Specific goals of the society are

- Promote research, education, and training in the environmental sciences.
- Promote the systematic application of all relevant scientific disciplines to the evaluation of chemical hazards.
- Participate in the scientific interpretation of issues concerned with hazard assessment and risk analysis.
- Support the development of ecologically acceptable practices and principles.
- Provide a forum (meetings and publications) for communication among professionals in government, business, academia, and other segments of society involved in the use, protection, and management of our environment.

These goals are pursued through the conduct of numerous activities, which include:

- Hold annual meetings with study and workshop sessions, platform and poster papers, and achievement and merit awards.
- Sponsor a monthly scientific journal, a newsletter, and special technical publications.
- Provide funds for education and training through the SETAC Scholarship/Fellowship Program.
- Organize and sponsor chapters to provide a forum for the presentation of scientific data and for the interchange and study of information about local concerns.
- Provide advice and counsel to technical and nontechnical persons through a number of standing and ad hoc committees.

SETAC membership currently is composed of more than 5000 individuals from government, academia, business, and public-interest groups with technical backgrounds in chemistry, toxicology, biology, ecology, atmospheric sciences, health sciences, earth sciences, and engineering.

If you have training in these or related disciplines and are engaged in the study, use, or management of environmental resources, SETAC can fulfill your professional affiliation needs.

All members receive a newsletter highlighting environmental topics and SETAC activities and reduced fees for the Annual Meeting and SETAC special publications.

All members except Students and Senior Active Members receive monthly issues of Environmental Toxicology and Chemistry (ET&C) and Integrated Environmental Assessment and Management (IEAM), peer-reviewed journals of the Society. Student and Senior Active Members may subscribe to the journal. Members may hold office and, with the Emeritus Members, constitute the voting membership.

If you desire further information, contact the appropriate SETAC Office.

1010 North 12th Avenue
Pensacola, Florida 32501-3367 USA
T 850 469 1500   F 850 469 9778
E setac@setac.org

Avenue de la Toison d'Or 67
B-1060 Brussels, Belgium
T 32 2 772 72 81   F 32 2 770 53 86
E setac@setaceu.org

www.setac.org
Environmental Quality Through Science®